Undergraduate Lecture Notes in Physics

Undergraduate Lecture Notes in Physics (ULNP) publishes authoritative texts covering topics throughout pure and applied physics. Each title in the series is suitable as a basis for undergraduate instruction, typically containing practice problems, worked examples, chapter summaries, and suggestions for further reading.

ULNP titles must provide at least one of the following:

- An exceptionally clear and concise treatment of a standard undergraduate subject.
- A solid undergraduate-level introduction to a graduate, advanced, or non-standard subject.
- A novel perspective or an unusual approach to teaching a subject.

ULNP especially encourages new, original, and idiosyncratic approaches to physics teaching at the undergraduate level.

The purpose of ULNP is to provide intriguing, absorbing books that will continue to be the reader's preferred reference throughout their academic career.

More information about this series at https://link.springer.com/bookseries/8917

Jorge Ernesto Horvath

High-Energy Astrophysics

A Primer

 Springer

Jorge Ernesto Horvath
IAG-USP, Astronomy Department
Universidade de São Paulo
São Paulo SP, Brazil

ISSN 2192-4791   ISSN 2192-4805 (electronic)
Undergraduate Lecture Notes in Physics
ISBN 978-3-030-92161-3   ISBN 978-3-030-92159-0 (eBook)
https://doi.org/10.1007/978-3-030-92159-0

Translation from the Portuguese language edition: *Astrofísica de Altas Energias: Uma Première* by Jorge Ernesto Horvath, © EDUSP 2020. Published by Editora da Universidade de Sao Paulo, Brazil. All Rights Reserved.

This Springer imprint is published by the registered company Springer Nature Switzerland AG
The registered company address is: Gewerbestrasse 11, 6330 Cham, Switzerland

Para las tres ovejas

Acknowledgements

My thanks to all the colleagues and students who have taught me physics and astrophysics over all these years, including Omar Benvenuto, José Antonio de Freitas Pacheco, Héctor Vucetich, Germán Lugones, Heman Mosquera Cuesta, Bete Dal Pino, Marcelo Allen, César Zen Vasconcellos, Adam Burrows, Márcio Catelan, Ignazio Bombaci, Rachid Ouyed, Ren-Xin Xu, Marco Limongi, S. O. Kepler, Mariano Méndez, F. C. Michel, Todd Thompson, Manuel Malheiro, Sérgio Barbosa Duarte, Genna Bisnovatyi-Kogan, Roberto Dell'Aglio Costa, Marcos Díaz, María Alejandra De Vito, Rodolfo Valentim, David Blaschke, Mark Alford, Paulo Sérgio Custódio, Dinah Moreira Allen, Thais Idiart, Gustavo Medina Tanco, Laura Paulucci, Márcio de Avellar, Rodrigo de Souza, Eduardo Janot Pacheco, Aurora Pérez Martínez, Daryel Manreza, Efrain Ferrer, Vivian de la Incera, J. C. N de Araújo, H. Stöcker, and W. Maciel among many others. I would also like to thank Antonio Lucas Bernardo and Lívia Silva Rocha, who served as teaching assistants for the AGA 315 High-Energy Astrophysics course and brought much to its success. Finally, the students of AGA 315 contributed to the development of this text in its final form. The Dean of Undergraduate Studies at USP and EDUSP are acknowledged for their patronage and production of the Portuguese version of this work. My daughter Katia Horvath is acknowledged for her dedication in correcting the uncountable number of typos and mistakes in the text, while those remaining are my responsibility entirely. The author also wishes to acknowledge the attention and assistance of Dr. Angela Lahee, Stephen Lyle, and the production staff at Springer for the completion of the present edition.

And yet, I still dream the dogmatic slumber of Immanuel Kant.

Contents

1   The Nature of the Physical World: Elementary Particles
and Interactions 1
  1.1   Elementary Particles and Fundamental Interactions:
         An Overview 1
  1.2   Elementary Interactions at High Energies 4
  1.3   Standard Model of Elementary Particles 8
  1.4   Strong Interactions and Quantum Chromodynamics (QCD) 9
  1.5   Gravitation as a Fundamental Interaction 13
  1.6   Role of Weak Interactions 14
  References 17

2   Elementary Processes at High Energies 19
  2.1   Genesis of the Concept of Photon 19
  2.2   Processes Involving Photons at High Energies (Absorption
         and Scattering) 21
         2.2.1   Photoelectric Effect 21
         2.2.2   Compton Scattering 23
         2.2.3   Pair Production 25
         2.2.4   Total Cross-Section and Absorption Coefficient
                per Unit Mass 27
  2.3   Relevant Processes for High-Energy Photons (Emission) 29
         2.3.1   Black Body Radiation 29
         2.3.2   Bremsstrahlung Radiation 30
         2.3.3   Synchrotron Radiation 34
         2.3.4   Čerenkov Radiation 38
         2.3.5   Sources of Positrons and Pair Annihilation
                (Emission Lines) 40
  References 43

3   Detection and Instrumentation in High-Energy Astrophysics   45
    3.1   Spatial, Spectral, and Temporal Domains   45
    3.2   CCDs in Optical Astronomy and High-Energy Astrophysics   48
    3.3   The Problem of Focusing (Imaging) High-Energy Photons
          and Its Solutions   50
    3.4   Space-Based Instruments for X-Ray and γ-Ray Detection   52
    References   55

4   Stellar Evolution up to the Final Stages   57
    4.1   Stellar Astrophysics   57
    4.2   Basic Facts and Observations   58
    4.3   Physical Description of Stellar Structure   61
    4.4   Some General Considerations about Stellar Evolution   71
    4.5   Stellar Evolution: Low Mass Stars   74
    4.6   Stellar Evolution: High Mass Stars   82
    References   87

5   Supernovae   89
    5.1   Supernova Types and Classification   89
    5.2   Supernovae and Gravitational Collapse (Types II, Ib, and Ic)   91
    5.3   Thermonuclear Supernovae   97
    5.4   Type Ia Supernovae and Cosmology   103
    5.5   Superluminous Supernovae   105
    5.6   Expansion of Supernova Remnants in the Interstellar
          Medium   108
    References   111

6   Astrophysics of Compact Objects   113
    6.1   Formation Events of Compact Objects: Statistics   113
    6.2   Theory and Observations of White Dwarfs   114
          6.2.1   In the Beginning...   114
          6.2.2   Matter in the High Density Regime
                  ($\rho \geq 10^3 \, \mathrm{g\,cm^{-3}}$)   116
          6.2.3   White Dwarf Structure   117
          6.2.4   Chandrasekhar Limit   120
          6.2.5   Observations of White Dwarfs   122
          6.2.6   Cooling and Crystallization of White Dwarfs   126
    6.3   Neutron Stars and Pulsars: Structure and Evolution   130
          6.3.1   The Pioneering Ideas   130
          6.3.2   Matter in the Neutronization Regime
                  ($\rho \geq 10^{11} \, \mathrm{g\,cm^{-3}}$)   131
          6.3.3   Relativistic Stellar Structure Equations (TOV)
                  and Neutron Stars   134
          6.3.4   Stellar Models and Comparisons
                  with Observations   136
          6.3.5   Pulsars and Other Neutron Stars   139

6.4 Physics and Observational Manifestations of Black Holes 145
6.4.1 Birth of the Black Hole Concept 145
6.4.2 What Do We Observe from Black Holes? 150
References 157

7 Accretion in Astrophysics 161
7.1 Roche's Problem 161
7.2 Spherical Mass Accretion and Accretion Disks 165
7.3 Binaries Containing Compact Objects: Observations
and Classification 170
7.3.1 Cataclysmic Variables (CV) 170
7.3.2 Low-Mass X-Ray Binaries (LMXB)
and High-Mass X-Ray Binaries (HMXB) 171
7.3.3 More on the Binary Systems Containing Black
Holes 173
References 175

8 Active Galactic Nuclei (AGNs) 177
8.1 Discovery of Quasars 177
8.2 Types of AGN and the Unified Model 181
8.3 AGNs and Structure Formation in the Universe 183
References 186

9 Neutrino Astrophysics 187
9.1 Neutrinos and Their Detection 187
9.2 Neutrino Sources: Solar Neutrinos 190
9.3 Neutrino Sources: Supernova 1987A 196
References 202

10 Gravitational Waves 203
10.1 Gravitational Radiation: The Basic Physics 203
10.2 Sources of Gravitational Waves 206
10.2.1 The Binary Pulsar PSR 1913+16 and Gravitational
Waves 208
10.3 Gravitational Wave Detectors 210
10.3.1 Interferometers and Resonant Masses: From
Dreams to Reality 210
10.4 Detection of Black Hole and Neutron Star Mergers: The
Beginning of a New Era 215
10.4.1 Overture: The Black Hole Merger Event
GW150914 215
10.4.2 The Aftermath: A Merger of Neutron Stars
in GW170817 217
References 220

11   Gamma-Ray Bursts 223
      11.1   The Problem of Gamma-Ray Bursts: The Most Distant
             Objects in the Universe? 223
      11.2   Models of the Bursts 227
      11.3   Recent Observations and Models of GRBs 230
      11.4   Fast Radio Bursts: A Related Phenomenon? 232
      References 235

12   Cosmic Rays 237
      12.1   Messengers from the Greatest Accelerators in the Universe:
             Cosmic Rays 237
             12.1.1   Origin, Propagation, and Acceleration 239
             12.1.2   Ultra-High Energy Regime 246
      References 255

Problems 257

Index 271

Chapter 1
The Nature of the Physical World: Elementary Particles and Interactions

1.1 Elementary Particles and Fundamental Interactions: An Overview

The idea that Nature is composed of discrete "packets" which combine to form the entire visible Universe originated in the classical Greek world. This idea of the elementary "granularity" of the physical world was implicit in Pythagoras' philosophy and his school in Crotone (now in Italy and then a part of the *Magna Graecia*) more than five centuries before the Christian era. The Pythagoreans attributed great importance to the discovery of the existence of a simple relationship between the tones of a string (whole numbers) and similar questions, thus foreseeing that the world was *discrete*, and formulated the powerful notion that reality is ultimately mathematical in nature. The modern version of elementary blocks (particles) and their interactions is presented in this Chapter.

Much more forcefully (although motivated by the logical solution to the problem of the illusory movement that had posed Parmenides, and not by any experimental evidence), the atomists Leucippus and Democritus formulated a theory regarding the nature of matter, where discrete units ($\alpha\tau o\mu o\zeta$, atom) moved in the absence of matter (vacuum), combining to produce the entire visible Universe. Atoms would differentiate themselves by their geometry (such as the difference between the figures "A" and "N"), by their disposition or order (such as the differences between "NA" or "AN"), and by their position (such as "N" is a rotated "Z"). Different combinations and proportions would be responsible for the diversity of bodies. This strongly materialistic doctrine (for example, for the atomists even the soul was made up of atoms) has never been fully accepted in general, and Aristotle and later philosophers raised objections against the atomic idea, which was almost totally forgotten. However, for many centuries this whole discussion did not go beyond the realm of ideas, since the technological development and methodological attitude of the Greeks never made a direct inquiry to Nature, and indeed the notion of experimen-

© The Author(s), under exclusive license to Springer Nature Switzerland AG 2022
J. E. Horvath, *High-Energy Astrophysics*, Undergraduate Lecture Notes in Physics,
https://doi.org/10.1007/978-3-030-92159-0_1

tal proof did not appear at all in the ancient world until at least the Low Middle Ages.

In spite of being far from accepted at the time of Leucippus and Democritus, versions of the atomistic world survived in Lucretius, and were quite influential in the exposition of the Epicurean version of the atomistic doctrine. His great work *De Rerum Natura* [1] even contains a hint of the idea of *inertia* applied to the motion of atoms, among other insights. However, the Aristotelian view of the world dominated the scene for almost two millennia, and little or no room was left for atomism. A first explicit break with Aristotle was due to the French clergyman Pierre Gassendi in the mid-17th century, resuming in many senses matters that had been addressed by the atomistic program [2]. Newton and others were able to formulate several ideas about the physical world that were reminiscent of the Greek Epicurean atomism.

Finally, in the 18th century atomic theory made a significant comeback, supported by the empirical work of John Dalton (1766–1844) and his followers. Dalton realized that the known chemical reactions were compatible with discrete packages and redefined the concept of the Greek atom for these, whereupon atoms began to have a tangible physical reality. Besides being firmly based on laws of conservation (for example, of mass), Dalton formulated the idea that a chemical reaction is basically a rearrangement of atoms. The complex substances are, in this vision, composed of atoms, in a very close parallel to the ideas of the Greek atomists. After a long debate and a lot of experimentation and argument, the "modern" atomic theory finally constituted a physical–chemical paradigm, anchored in the "new atomism" of Newton, Boltzmann, and others, for which Dalton's insistence and empirical evidence-gathering were truly fundamental.

Nevertheless, it took almost another century for atoms to begin to show their true nature. In fact, in the original version, and even by the etymology itself, atoms remained indivisible. However, the first important experimental milestone confirmation of the works of Faraday and others on electromagnetic theory, together with the ideas of Boltzmann and Gibbs who sought to ground Thermodynamics in microphysics, came with the discovery of the electron by J.J. Thompson in 1897. Shortly afterwards, Rutherford conducted a series of experiments that revealed the existence of the atomic nucleus; and in the early 20th century, the discovery and interpretation of radioactivity in terms of the structure of the nucleus led to the identification of the proton and neutron as the basic ingredients of Rutherford's nucleus.

With these discoveries began an era of characterization of the particles that constitute atoms (that is, there was a change to a deeper level of elementarity) and construction of detailed atomic models. In fact, one of the first models (the Thompson model) postulated a positively charged (continuous) fluid in which the electrons were embedded, that is, the amount of charge assigned to each component was not defined. It was up to Robert Millikan to contribute to the problem shortly afterwards with his experiments, which demonstrated the discrete character (quantization) of the electric charge of the electron.

From both the theoretical and experimental points of view, the concept of *elementarity* can be considered in a relative sense: a particle can be seen as elementary (without internal structure) at low energies, but can reveal itself to be composed of

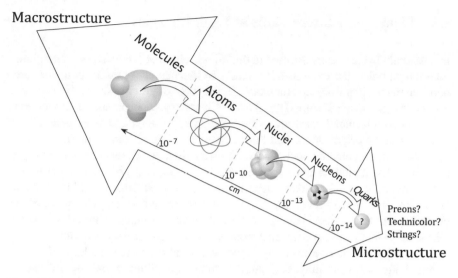

Fig. 1.1 The structure of matter. As higher energies are attained, matter reveals its most elementary components at the smallest scales. So far there is no evidence of any substructure for quarks or electrons, and nothing has yet been detected even at distances $\sim 10^{-15}$ cm, although some theoretical proposals have been formulated, such as those indicated in the figure with a question mark

something at higher energies. To find out whether a particle is composite or truly elementary (at least up to a definite energy scale), we have long employed a method that uses brute force: if thrown against known targets, the particle will reveal a substructure as long as the energy delivered is sufficient (for example, throwing a nut against a wall will reveal its internal structure if the energy is sufficient to break the shell). Thus, in collisions between protons performed in particle accelerators, these appear to be particles without internal structure as long as the collision energy is low. In collisions at very high energies carried out in large accelerators, protons show that they are composed of other, more fundamental particles, the quarks. In this model, three quarks make up a proton or neutron in such a way as to comply with the basic observed facts. We will soon have more to say about these quarks.

The limits to the *spatial resolution* we can reach correspond to the limits of elementarity that we can explore, and the search for the most fundamental theory amounts to attempting to find the absolute limit for the composition of matter, not just what can be measured in a given experiment. As experiments have accelerated particles to ever larger energies, we have discovered ever smaller structures. In fact physicists today are able to measure distances of up to one-hundredth of a *fermi*, or 10^{-15} cm, the present limit of elementarity. The known hierarchy of the composition of matter is shown in Fig. 1.1. The question here is whether elementarity effectively has an absolute limit, or whether an increase in the energy of the collision will necessarily always reveal a new underlying structure (an interesting perspective on this issue can be found in David Bohm's work [3]).

1.2 Elementary Interactions at High Energies

In Classical Physics we are not used to thinking in terms of "elementary interactions" between particles, but in terms of "forces". In Mechanics we think about the force between macroscopic objects; for example, the force of gravity between two masses m_1 and m_2, given by $-Gm_1m_2/r^2$. Also in Chemistry, concepts such as interatomic forces, intermolecular forces, and so on are regularly used, these being generally of electromagnetic origin. As is well known, almost all these forces are, in principle, derivable from a potential, and express the way elements of matter attract or repel each other. Such pictures have a strong classical foundation, of mechanical origin, but an important question today concerns the way we should understand interactions at the most elementary level, that is, among the elementary particles themselves, going beyond the classical concept applicable to macroscopic "chunks" of matter, which are in fact made up of an enormous number of elementary particles.

In the micro-world of elementary particles most of the notions we have of macroscopic matter fail ostensibly. It is not that there is anything wrong with Classical Physics; on the contrary, for several centuries many aspects of the physical world have been successfully explored using those ideas. But it would be wishful thinking to expect concepts developed in the classical world to apply as-is to the micro-world, without modification. In fact, this kind of extrapolation has given rise to countless problems and "paradoxes" that still plague description of the elementary world. The development of Quantum Mechanics in the 20th century exposed many of these contradictions without really solving their exact nature, since the so-called *interpretation* of the quantum formalism required to make sense of the ideas remains unsatisfactory and is still a subject of discussion and research. This idea of "interpretation" is consensual in other cases (for example, Classical Mechanics), where the meaning of the relevant concepts and their role in the physical description is unambiguous. This is not the case with the quantum formalism. A discussion of this problem would take us too far from our objective here, and we only mention this situation in passing (see [4] for an in-depth discussion of these problems).

Although physicists do not yet have a complete clarification of the interpretation of Quantum Mechanics (QM), it is notable that each time the theory is required to provide a quantitative (probabilistic) prediction regarding an experiment, it provides values that are in good agreement with the measurements (!). One of the characteristics of QM that is surely common to any interpretation, and that constitutes a breaking point with Classical Physics, is provided by the so-called *uncertainty relations*. This concept is important for the rest of our discussion and will be described below.

We are familiar with the existence of errors in the measurements of any classical theory. For example, if we measure the position of a moving test particle, the height of a mountain, or any other spatial variable x_0, there will necessarily be an error associated with each measurement, say δx. To reduce this error we can measure repeatedly, improving the determination of the position of the object x_0, and calculate the statistical error Δx, which is the so-called standard deviation, assuming that there are no systematic errors affecting the measurement. As a large enough number of measurements is accumulated, the error $\Delta x \to 0$, that is, there is in principle no obstacle to reducing it indefinitely by making more and more measurements.

In QM the situation is radically different. Variables such as the position x_0 and the linear momentum p_0 of an object, referred to as *conjugate quantities*, cannot be measured simultaneously with arbitrarily small errors. The product of the errors in the two variables is always greater than the action quantum $\hbar = h/2\pi$, where h is Planck's constant. The quantum object does not behave like any classical equivalent, and cannot be located with increasing precision without its momentum acquiring increasing uncertainty, or vice versa. In summary, we have

$$\Delta x \times \Delta p \geq \hbar, \tag{1.1}$$

which are known as the *uncertainty relations* (although a more exact translation of the original German word used by Heisenberg would be "indeterminability"). Similarly the energy E measured at a time t should satisfy

$$\Delta E \times \Delta t \geq \hbar, \tag{1.2}$$

although this last inequality has roots in Classical Physics, since it does not stem from the algebra of operators like (1.1).

Using such uncertainty relations, we are now in a position to elaborate some of the basic ideas that we will use later in the course. The first concerns the possibility of *violating energy conservation* in a process, although only for a very short time. In fact, in the quantum vacuum, (1.2) allows particle–antiparticle pairs to "pop up" spontaneously and quickly annihilate again (Fig. 1.2), as long as the time elapsed is short enough. This phenomenon is known as *vacuum fluctuation*, and is unique to quantum theory. The particles involved are said to be *virtual particles*.

A particle of this type with energy E and mass m can be exchanged for two "real" particles up to distances $L \approx c \Delta t$ (the speed of light c is the maximum speed of the

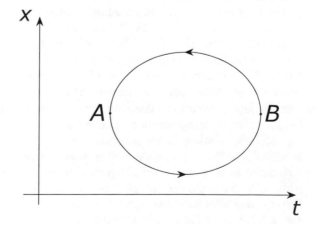

Fig. 1.2 Quantum vacuum fluctuations. At point A, a particle–antiparticle pair appears with total energy ΔE, then annihilates at point B after a time $t(A \to B) \leq \hbar/\Delta E$. This pair is said to be *virtual*, and does not compromise the conservation of energy at the macroscopic level, since it "vanishes" in a very short time, under the threshold of the uncertainty limit given by (1.2)

Fig. 1.3 The range of a quantum interaction. The two players exchange a virtual ball (particle) of mass m. This exchange is possible for successively shorter distances as m grows. In the limit $m \to 0$, the interaction has infinite range, that is, the "players" can be as far apart as they want

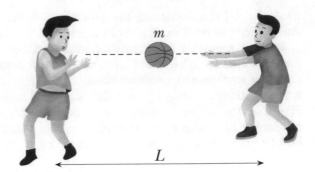

virtual particle). In terms of mass, we can thus write

$$L \approx c\Delta t = \frac{\hbar}{mc} \, , \tag{1.3}$$

where we have used the famous relation $E = mc^2$. The quantity $\lambda_C = \hbar/mc$ with length dimensions on the right-hand side is called the *Compton wavelength*, and allows a simple interpretation of the physical situation: the intermediate particle of mass m is located in a region of linear dimension roughly the order of λ_C where the probability of finding it is not zero. Thus, the Compton wavelength defines a range for the interaction mediated by a particle of this mass. When we consider particles of higher mass, λ_C decreases and the interaction has a shorter range. A classical analogy for this behavior is presented in Fig. 1.3.

Although the detailed calculation of each elementary process is complicated, requiring knowledge of the quantum theory of fields, some order-of-magnitude estimates can be obtained using simple dimensional analysis. The modern view of fundamental interactions is that these are essentially virtual intermediate particle exchanges between two or more particles that carry some form of charge (for example, the electron with an electric charge).

The long range of the electromagnetic interaction suggests that the mediating particle has no mass, and that it is the quantum of "light", i.e., the *photon*. The charge associated with the interaction in this case is just the electric charge, and the strength of the interaction is measured in terms of some dimensionless number, which in the case of Electromagnetism is the *fine structure constant* $\alpha = 1/137$. This coupling constant is small and indicates that the electromagnetic interaction is much weaker than the strong interactions that bind the nucleus, for which the coupling constant has the value ≈ 1 in appropriate units. As we have seen, besides the exchange of the mediating particle, fluctuations can occur thanks to the uncertainty relations, in which a particle–antiparticle pair or more complex entities can appear and get reabsorbed in very short times. Representing the electrons (or any other charged particle) by a straight line and photons by wavy lines, we can draw "hieroglyphs" that represent what happens diagrammatically in each possible case (Fig. 1.4). These symbols correspond to definite mathematical expressions in each case, but we will

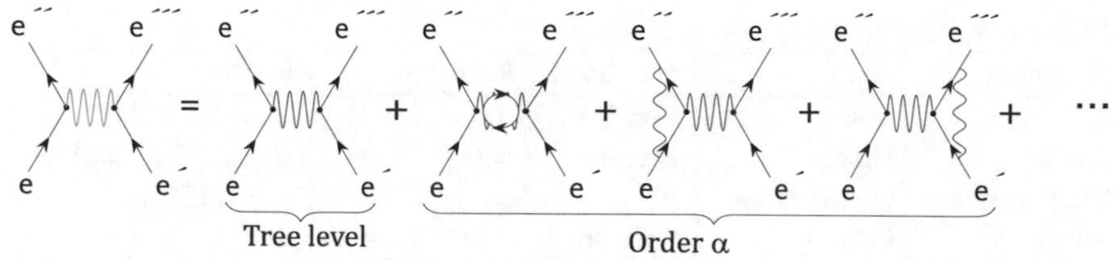

Fig. 1.4 Basic electromagnetic interaction between two electrons as a sum of quantum processes which are progressively less important

not discuss them here because we are only interested in the conceptual aspect of these calculations.

The result is an infinite sum of such diagrams that takes into account all the complexity of the micro-world, and from which the known classical limit emerges under certain conditions. In Fig. 1.4 the diagram on the left should be considered as including all the interactions plotted on the right, and is called the "dressed" diagram. On the right, we see that the first and simplest diagram does not contain vacuum contributions; these appear only in the next order in powers of the coupling constant α. Thus, the classical theory of the electromagnetic field due to Faraday and Maxwell corresponds to the first (classical) diagram, referred to in the jargon as the *tree level* (no fluctuations included). The quantum corrections appear at order α and higher powers of α, whence the importance of having $\alpha < 1$: although the series cannot be summed in general (in fact, it is not a convergent series in the mathematical sense, but is nevertheless increasingly accurate as more terms are added!), the terms are ever smaller as the power of α increases, so the result can be found as accurately as one requires by calculating the contributions up to a given order. Taking the static limit of the series and staying only with the tree level, we recover the classical Coulomb and Yukawa potentials, viz., $V_C(r) \propto -\alpha/r$ and $V_{Yuk}(r) \propto \exp(-r/L)$, respectively, in the case where the mediator has zero mass or mass m [see (1.3)].

The construction of a coherent picture of the physical Universe needed the recognition of the existence of four elementary interactions: the electromagnetic interaction already mentioned, gravitation (still without a proper quantum theory), the strong interaction (responsible for nuclear binding), and the weak interaction (responsible for beta decay and other processes involving neutrinos). The same particle can have more than one "charge" and thus suffer several of these interactions. This is the case, for example, for the quarks inside a proton, which have electric charge, but also weak charge and strong charge (called "color"). Thus, they are capable of interacting by exchanging photons, gluons, and massive bosons. A summary of the fundamental interactions and their main characteristics is shown in Table 1.1.

Obviously, it is not always the best strategy to try to understand a certain physical problem directly in terms of the four fundamental interactions, since the problem can sometimes get very complicated. Thus, we are again led to consider the *absolute* or *relative* concept of elementarity, depending on the physical conditions involved and the need for detail in the description, as already pointed out. As a concrete example of

Table 1.1 Summary of elementary interactions

Interaction	Charge	Mediator	Range	Intensity
Gravitation	Mass	Graviton	∞	$\alpha_G = 5.9 \times 10^{-39}$
Weak	Flavor	$W^{\perp}, Z^0$	$\approx m_W^{-1}$	$G_{\text{Fermi}} = 10^{-5} m_p^{-1}$
Electromagnetic	Electric charge	γ	∞	$\alpha = 1/137$
Strong	Color	8 gluons	$\approx m_\pi^{-1} \approx 1.5$ fm	$\alpha_S \approx 1$

this situation, the strong interactions (those that hold nuclei together) can be described as mediated by pions and mesons, as in the 1950s and 60s, as long as the energy considered is low (these are considered "effective" descriptions in an ample sense). If we increase the energy, the protons and neutrons and also the mediating particles will reveal their composite nature and that simple description may be insufficient. Note once again that this is *not* a statement about the absolute elementarity of the electron, quarks, or other particles, but rather one about the relative elementarity for practical purposes.

1.3   Standard Model of Elementary Particles

Based on the previous ideas we can now discuss the known "zoo" of particles, which make up the so-called Standard Model of particle Physics, and classify the elementary interactions. The world of subatomic particles has expanded vertiginously since the early 20th century. In the first decades of that century, only the electron and the proton were known. The discovery of the neutron and soon after antiparticles brought considerable perplexity and great challenges for physicists studying the structure of matter. The first particle accelerators idealized and built by E. Lawrence in the United States gave a further stimulus to the Physics of elementary particles, boosting the discovery of new particles by increasing the energies involved in collisions.

In addition to modelling the dynamics of interactions between these particles, a classification scheme was also required. After several attempts of historical interest, but whose complexity would take us too far from the scope of the present book, there is a consensus today around the scheme that became known as the *Standard Model*. This classifies elementary particles into three groups or generations, depending on their participation in the elementary processes that have been detected, i.e., the reactions in which the particles take part. The composition of a generation is always the same: it contains two quarks (which constitute the baryons and mesons), a charged lepton (the electron, the muon, and the tau, successively), and a neutrino associated with the latter (a different neutrino for each type of lepton). The discovery and identification of these particles, and the recognition of the symmetries implemented over the generations, took several decades and was only completed with the discovery of the quark t in 1995 and the Higgs boson (responsible for the observed masses) in 2012. At the present time there is no evidence to indicate any important departures from the Standard

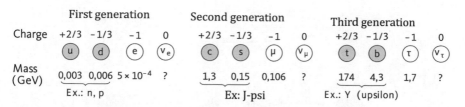

Fig. 1.5 Contents of the Standard Model. The figure shows the three generations with the two quarks (*up* and *down* in the first, *charm* and *strange* in the second, and *top* and *bottom* in the third, all fantasy names devised to make them easy to remember. Also shown are the charged leptons e, μ, and τ and the three corresponding neutrinos ν_e, ν_μ, and ν_τ. Electric charges and masses of the particles are indicated above and below each particle. See [5] for more detail

Model data. Figure 1.5 illustrates our current knowledge of the particle structure of the Standard Model.

1.4 Strong Interactions and Quantum Chromodynamics (QCD)

In the early twentieth century, the recognition of the need for a new force to hold the atomic nucleus together led to the introduction of the Yukawa potential, as already mentioned, and to the prediction of the existence of the pion, discovered soon after as a component of cosmic rays and in dedicated experiments carried out by the physicist César Lattes and collaborators at the University of São Paulo in 1947 (see his account in [6]). This exemplifies the idea that interactions are the result of the exchange of mediating particles. Later on in nuclear Physics it became clear that the pion was only one such mediating particle. The interactions between nucleons (protons and neutrons) also involve the exchange of kaons, ρ mesons, and other mediators, giving rise in the static limit to the so-called Yukawa potential and corrections presented above.

For some decades this picture was satisfactory, but accelerator experiments eventually showed that protons and neutrons were far from being pointlike: incident electrons striking these nuclei scattered as if they encountered "hard" points on scales $\leq 10^{-14}$ cm. Thus, Gell-Mann and Zweig were led to suggest that there are fundamental constituents of the nuclei, which they called quarks. A highly non-linear field theory called quantum chromodynamics (QCD) was soon developed, in which quarks exchange gluons, the mediating particle of strong interactions. The associated charge comes in three types and was fancifully called color (although it has nothing to do with real colors, of course). However, this theory has a characteristic that really sets it apart: despite intensive searches it has never been possible to detect an isolated quark outside a hadron (hadrons are particles participating in strong interactions, i.e.,

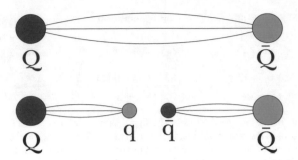

Fig. 1.6 Color confinement. A meson formed by a quark Q and an antiquark $\bar{Q}$ receives enough energy to break up, but the gluon "string" uses this energy to create a quark q and an antiquark $\bar{q}$, and thus two "white" mesons are formed, in which the colors neutralize one another, thus keeping the total color invisible to an external observer

baryons, such as the nucleons, and mesons). This gave rise to a totally new idea, that of the *confinement* of color, according to which the colors of the quarks always combine (just like the primary colors) to produce a "white" hadron, that is, without color. Each time a quark is ripped out of a hadron, this breaks the flow tube that connects it with another, and thus two mesons are produced (Fig. 1.6).

It is currently believed that this property of quarks (and gluons) is contained in the theoretical description, since there are numerical simulations that demonstrate confinement. But there is also another peculiarity of the theory: at very short distances, inside the hadrons, the quarks and gluons seem to be *free*, that is, they do not "feel" the interactions between them. In fact, we can define these short distances or long distances by using the relativistic definition of the relation between momentum and energy, i.e., $E = pc$, of the incident particle. From the uncertainty relation (1.1), we have immediately that the distances reached by the projectile particle are inversely proportional to its energy, $x \sim \hbar/E$. The "long" distances can be considered as those greater than the radius of the proton, while the small ones are much smaller than this radius. The behavior of the quarks in the first case is called *infrared slavery* (low incident energies) and in the second *asymptotic freedom* (high incident energies).

This behavior can be simulated by the phenomenological potential

$$V(r) = -\alpha_S/r + kr .$$

At short distances the first, the attractive term dominates, but if we consider large distances the potential grows and it will be impossible to extract a quark. Another widely used approach, which we will describe below, draws a physically reasonable picture and is simple to calculate, whence it has been widely used. It is known as the *MIT bag model* (see Fig. 1.7).

To simulate the effects of confinement, the model admits that the true vacuum is a perfect dielectric for the color charge, and particles that carry color do not penetrate it. Thus, the model proposes to consider a "perturbative vacuum" cavity in the true vacuum where quarks and gluons can live. Energy is required to create this cavity

Fig. 1.7 The MIT bag as a bubble in the true QCD vacuum, characterized by a (constant) negative pressure $-B$. It is within this cavity that the quarks roam freely, with a kinetic energy that helps balance the configuration and stabilize the bag

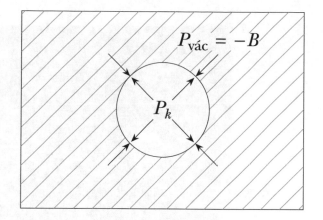

(bubble), and its density is represented by a constant value B. There is a pressure inside the cavity (it is a bubble) equal to $P_{\text{vac}} = -B$. The pressure balance required for equilibrium is produced by the pressure of the particles "living" inside the P_k cavity. Now, as we are talking about elementary quarks, the states of energy that can exist in the cavity are discrete, that is, they are quantized. The complete calculation, which involves solving Dirac's equation in a spherical well and exceeds the scope of this discussion, shows that for N quarks the energy is $E_k = 2.04N/R$, where R is the radius of the assumed spherical cavity and the mass of the quarks is neglected. The total energy is

$$E_{\text{tot}} = E_{\text{vac}} + E_k = \frac{4}{3}\pi R^3 B + \frac{2.04N}{R} \,. \tag{1.4}$$

The equilibrium configuration must be a minimum of the total energy, found by varying it against the radius R:

$$\frac{\partial E_{\text{tot}}}{\partial R} = -\frac{2.04N}{R^2} + 4\pi R^2 B = 0 \,, \tag{1.5}$$

a condition that determines the cavity radius to be

$$R = \left(\frac{2.04N}{4\pi}\right)^{1/4} \frac{1}{B^{1/4}} \,. \tag{1.6}$$

Setting $N = 3$ and inferring the value of B by means of a fit procedure to the known hadron masses, the value of the nucleon (proton or neutron) radius $R_N = 1.13$ fm $[B/(145 \text{ MeV}^4)]^{(-1/4)}$ will emerge, since these correspond to the "bubbles" containing three light quarks. The full model is much more complex than we have explained here, but this simple calculation serves to understand the basics of the constitution of hadrons in terms of QCD (or, more precisely, a phenomenological model based on it).

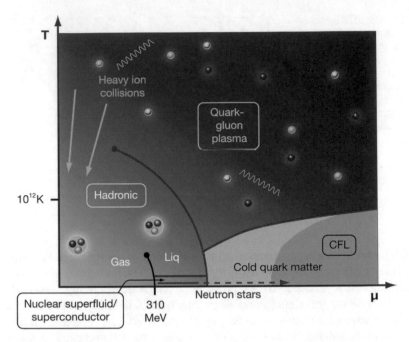

Fig. 1.8 Basic phase diagram of QCD. The QGP sits in the *upper and right regions*, accessible in heavy-ion collisions (*arrow trajectories*), the early Universe (along the *vertical axis*), and neutron stars (cold quark matter), as indicated. Courtesy of Mark Alford and Alan Stonebraker, Washington University

For applications to the early Universe, a statistical treatment of QCD considering a very large number of quarks and gluons has shown that this confined phase should give rise to a so-called quark–gluon plasma (QGP) above a temperature of some 170 MeV (approximately 2×10^{12} K). The nature of this transition is being debated, but it seems to be a rather logical consequence of asymptotic freedom. According to these estimates, the primordial Universe confined color some 10^{-5} s after the Big Bang. For the same reason (asymptotic freedom), it is expected that ordinary hadrons will dissolve into quarks in the regime of low temperature and high density. The question here is whether density is relevant to the interiors of neutron stars, a natural "laboratory" where the enormous gravitational field would cause this transition by "squeezing" protons and neutrons. The situation is summarized in Fig. 1.8.

In short, the truly fundamental constituents of hadrons, the quarks, were revealed by scattering experiments in the 1970s. The theory that describes them, QCD, is difficult to solve, but it seems to contain the mathematical elements that lead to the observed physical behavior. Moreover, QGP has been observed in heavy ion collisions, while there is a consensus around the existence of this phase in the primordial Universe and it may well be present in the interiors of neutron stars. The quark model indicates that the previously known nuclear interactions are what is "left over" from integrating (summing) over the most fundamental degrees of freedom of QCD. Each of the contributions associated with a meson mediator responsible for interaction, in the original vision of nuclear Physics, then needs to be understood in terms of quarks and gluons, a complex task that is currently underway.

1.5 Gravitation as a Fundamental Interaction

Many textbooks begin with a discussion of the classical gravitational force between two macroscopic masses m_1 and m_2

$$F_G = -G_N \frac{m_1 m_2}{r^2} , \tag{1.7}$$

where G_N is Newton's gravitational constant and r the distance between the masses. With a view to formulating the description of gravitation as an elementary interaction, where the particles exchange a "graviton" (Table 1.1) and from which the law of Newtonian gravitation should result, dimensional analysis of Newton's equation shows that G_N is not dimensionless, a fundamental requirement for the construction of an elementary theory. That is why a "gravitational fine structure constant" is commonly defined by $\alpha_G = G_N m_p^2 / \hbar c$ on an energy scale equal to the mass of the proton (the quantity that appears in Table 1.1). But by the tiny numerical value of $\alpha_G \sim 10^{-38}$, gravitation can almost always be ignored compared to the other forces of Nature, at least as long as we talk about elementary processes. The question that arises is: why is it then that gravitation dominates the structure of the observable Universe, stars, and galaxies? The simplest answer is to be found in the unique nature of the "charge" of the gravitational field, which is just mass: if macroscopic sets of particles are considered, gravitation "accumulates" until the structure itself is dominated by it, while the other forces cancel each other out as we consider more and more particles. Let us consider quantitatively N particles of equal mass. The radius of a sphere formed by this set of particles depends on $N^{1/3}$, while the energy of the gravitational bond is proportional to $N^{2/3}$. To compensate for the smallness of the factor of 10^{-38} of the constant α_G, the number of particles required must be $N = 10^{38 \times (3/2)} = 10^{57}$. This is approximately the number of particles (protons) in a star like our own, with mass denoted by $M_\odot$, and results in the "natural" scale where gravitation becomes more important than the other forces at a macroscopic scale (in fact we know that the Sun, for example, does not have a large contribution to its binding energy from strong, weak, and electromagnetic interactions) [7].

This discussion leads to the conclusion that we can neglect gravitation in microscopic systems, unless the energy scale grows as much as to make $\alpha_G \approx 1$. Under these conditions, microscopic gravitation would be as important as the other fundamental interactions. The mass where this equivalence occurs is

$$m_{Pl} = \left(\frac{\hbar c}{G_N} \right)^{1/2} , \tag{1.8}$$

the so-called *Planck mass*, with associated energy $E_{Pl} = m_{Pl} \times c^2 = 10^{19}$ GeV. As the most energetic phenomena in the laboratory, and even in the extreme cosmic rays of ultra-high energy discussed in Chap. 12, are still many orders of magnitude below this value, we will never have to worry about gravitation as an elementary theory, i.e., its quantum version. This is fortunate, since we do not yet have a consistent theory of quantum gravitation. Although the basic contribution should be the exchange diagram of an intermediate particle (or graviton) between any two massive particles, no quantum calculation is fully consistent. On the other hand, the classical versions of Newtonian gravitation and General Relativity have had spectacular success. Although we would like to have a quantum theory of gravitation, it has never been possible to build an acceptable version. When proceeding in the same way as in the quantization of other field theories, there is a divergence of the quantum theory of gravitation above a certain order in standard perturbation theory. Many physicists believe that there is a strong analogy here with the history of weak interactions, since Fermi's quantized theory also leads to divergent results beyond a certain order in perturbation theory. It may be that Einstein's theory of gravitation is not a fundamental theory, but only an "effective" theory, akin to the Fermi case. Thus, physicists still live in a dual world where they know that, on the one hand, the microscopic world is described by the laws of Quantum Mechanics, and on the other, gravitation behaves in a classical way as far as we can measure and observe, and these two descriptions are incompatible. The solution of this antagonism is what motivates the search for unified theories.

1.6 Role of Weak Interactions

In the 19th century, thanks to contributions from Maxwell, Faraday, and others, Electromagnetism was established as a theoretical paradigm for the study of phenomena involving electric charges in the laboratory. The discovery of the electron by J.J. Thompson in 1897 (the quantum of electric charge *par excellence*) provided a way to "penetrate" the atom by throwing electrons at it, and later to discover the atomic nucleus using helium nuclei (also electrically charged) as projectiles. The observation of the behavior and composition of atomic nuclei then opened an important window in the study of elementary particles.

By the 1920s, the proton had been identified as a component of the Rutherford nucleus. A series of experiments showed that, under certain circumstances, a nucleus could change its state of charge, with the expulsion of an electron from the nucleus. Thus, there were two possibilities: either the atomic nucleus contained electrons, or they were emitted by a particle decaying into a proton and an electron. This last hypothesis received definitive confirmation when Chadwick discovered the neutron in 1931. It was found that neutrons could spontaneously convert into protons, either when free or within the nucleus, whence Nature could change the type of nucleon that constituted the nucleus under certain conditions.

It also became clear that the observed conversion was not of electromagnetic origin (although the electric charge was conserved). Physicists thus sought the origin and nature of the force responsible. In the first place, it had to be a short-range force because the reaction takes place mainly on scales of the order of the atomic nucleus. The characterization of the strength of this force also emerged from the data, and turned out to be several orders of magnitude weaker than the electromagnetic force (see Table 1.1). Thus, the discovery of weak forces associated *neutron decay* with a new fundamental interaction:

$$ n \rightarrow p + e^- + \bar{v}_e \, , \tag{1.9} $$

where the neutron and proton were still part of the nucleus, and the electron escaped from the nuclear region. The last protagonist here, in fact an anti-neutrino, was not observed at first, but was postulated by W. Pauli to solve two serious problems with this decay: the conservation of energy and the conservation of angular momentum in the reaction. In fact, in spontaneous decay, such as was observed for neutrons within nuclei, the total angular momentum did not seem to be conserved, since the neutron spin $(1/2)$ was equal to half the spin of the particles observed in the reaction products, a proton of spin $1/2$ and an electron of spin $1/2$. Moreover, the sum of the energies of the particles taking part in the reaction was not constant. Nobody liked to abandon the conservation of energy and the angular momentum in Physics, and this is what inspired Pauli's creative solution to this problem.

In fact, he postulated a neutral particle that had to be very light or of zero mass, with the necessary spin $(1/2)$, to restore the conservation of both quantities. Basically, one could describe this hypothesis as the emission of a spin quantum. The important thing was to restore the conservation laws.

The elementary theoretical description of decay (corresponding to the simplest theory) was formulated by Fermi from 1933, and can be visualized in terms of Feynman diagrams. The basic diagram is the one shown in Fig. 1.9, which corresponds to the reaction in (1.9). The formal mathematical description incorporated a new term that led to the violation of parity, that is, the invariance of the processes under a change of coordinates $(x, y, z) \rightarrow (-x, -y, -z)$. The discovery that some weak interactions are not the same when "viewed in the mirror" (such is the meaning of the above sign change transformation) was confirmed experimentally in the 1950s by observing specific particle decays, and it led to the 1957 Nobel Prize for T.D. Lee and C.N. Yang, although it was C.S. Wu who conducted the crucial experiment that positively demonstrated this effect [5].

The reverse decay process $p + e^- \rightarrow n + v_e$, which involves emission of a neutrino, happens under physical conditions in which electrons can be captured by protons with energy gain (for example, near the end of the life of a star, see Chap. 4). Fermi's theory is oversimplified, but sufficient for low energies, and it is still used today in these cases. When *gauge theories* emerged, with the theoretical work of the 1960s and after, it became clear to physicists that Fermi's theory was actually a simplified version of one of them. Remembering that, for low energies, the spatial and energy "resolution" of the experiment is not sufficient (or equivalently that the mediator is much more massive than the energy of the measurement), we conclude that the decay shown in Fig. 1.9 can be thought of effectively as a diagram where, in

Fig. 1.9 The most important decay diagram in Fermi's theory. Note that it is simpler than the one in Fig. 1.4, since it seems to ignore the propagation of a mediating boson, which happens "inside the point". This version is valid if the energy is low enough

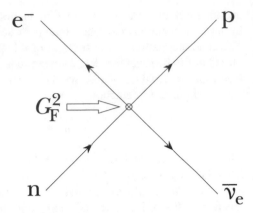

the central "point", a mediator ($W^\pm$ or Z^0 in Table 1.1) is emitted and then decays into the final pair. With this idea, we see that the emission and reabsorption vertices coincide at this central "point", whence Fermi's theory works at very low energies. However, it should be remembered that it is an approximation. The mass of the Z^0 boson determined at CERN in the 1980s is approximately 90 GeV, and this value justifies *a posteriori* the original data regarding the range of the weak force [see Fig. 1.3 and (1.3)]. Using the uncertainty relation $\Delta E \times \Delta t \geq \hbar$ and inserting the value of the mass (energy) of the Z^0, we obtain a limit for the average lifetime of 3×10^{-25} s, which corresponds to a maximum range for the interaction of $\Delta t \times c = 10^{-14}$ cm, i.e., less than the radius of a nucleon.

The progress of research in the 1970–80s led to the consideration of a variety of weak reactions of astrophysical (and cosmological) interest. However, in all of them, the tiny value of the cross-section, a direct consequence of the small value of the Fermi constant in Table 1.1, implies that high temperatures and/or densities are required for the weak interactions to be important for large particle sets. Above a certain scale, Fermi's theory is no longer valid and needs to be replaced by the corresponding expressions of the *Salam–Weinberg model*, the gauge theory developed for these purposes, in which symmetry breaking plays a fundamental role in obtaining the trio of massive bosons ($W^\pm$, Z^0) that lead to weak interactions, while the photon γ remains massless, as it should to mediate the electromagnetic interactions.

We have already said that, since their discovery, weak interactions have posed serious problems to physicists. For example, Pauli's introduction of the neutrino was a bold hypothesis, but in the end proved to be correct. Due to its "ghostly" nature, it was very difficult to study and characterize the neutrino experimentally. Spin and momentum (energy) are their only two characteristics, and there are only two dynamic possibilities: either the spin **s** is opposed to the direction of the momentum **k**, or it is in the same direction (Fig. 1.10). These two cases correspond to the particle and antiparticle, respectively, since they cannot convert one into the other in the absence of mass. Thus, in the first case, the particle was called a *neutrino* and in the second, an *anti-neutrino*, both names being due to Enrico Fermi, who used the diminutives of "neutron" in Italian when he coined the terms.

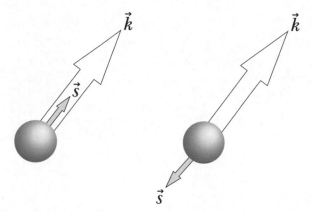

Fig. 1.10 Neutrinos and antineutrinos. If their mass is zero, neutrinos and antineutrinos cannot be confused, since the former always have their spin in the opposite direction to their momentum (*right*), while the latter always have their spin in the same direction as their momentum (*left*). Oscillations between the two types are possible when there is a (small) mass, and this has been identified as the cause of the solar neutrino problem, discussed in Chap. 9

It is important to point out that the relative direction of the spin and momentum is an affirmation of "absolute" character, since the projection (called the *helicity*) is invariant under conjugation of charge followed by parity inversion (a CP transformation). A zero-mass neutrino is described by a two-component function called a *spinor*, but this is not so for finite mass particles. We still do not know whether massive neutrinos require two or four components for their description (although it is agreed that their mass is *not* zero, because of the consistent detection of fewer solar neutrinos than expected over the years). We will see in Chap. 9 how it has been possible to develop neutrino Astrophysics from a basic knowledge of the cross-sections and source fluxes, while the questions of mass now revealed by oscillations still holds the attention of the community today.

References

1. Lucretius, *Lucretius the Way Things Are: De Rerum Natura*, translated by Rolfe Humphries (Indiana University Press, Indianapolis, 1968)
2. P. Gassendi, *Stanford Encyclopedia of Philosophy*. https://plato.stanford.edu/entries/gassendi/
3. D. Bohm, *Wholeness and the Implicate Order*, 1st edn. (Routledge, New York, 2002)
4. M. Bunge, *Philosophy of Physics* (Dordrecht-Holland, Boston, 1973)
5. T.D. Lee, *Symmetries, Asymmetries, and the World of Particles (Jessie & John Danz Lectures)* (University Washington Press, Washington, 1987)
6. C.M.G. Lattes, My work in meson Physics with nuclear emulsions, in *1st International Symposium on the History of Particle Physics*, eds. J. Bellandi Filho and A. Pemmaraju. Topics on Cosmic Rays, 1, pp. 1–5 (1981)
7. A.S. Burrows, J.P. Ostriker, Astronomical reach of fundamental Physics. Proc. Natl. Academy Sci. **111**, 2409 (2014)

Chapter 2
Elementary Processes at High Energies

2.1 Genesis of the Concept of Photon

The long and fascinating history of the study of light and electromagnetic phenomena (unified through Maxwell's equations) went through several stages before we reached the contemporary view. Without going into the details of earlier thought, an important founding contribution at the beginning of the 18th century was the publication of Newton's ideas in his book *Opticks* (1704), where Newton challenged the accepted view of the nature of light which went back to Aristotle's time, thus laying the foundations for extensive further debate. Newton defended a mechanistic framework, arguing that light was composed of material corpuscles, basing his scientific deductions on a series of experiments, including his famous example of the chromatic decomposition of light when it passes through a prism. Elementary scattering, absorption and emission processes involving photons are presented with an eye for their application in High-Energy Astrophysics.

This work had great impact, and led the way to important developments. In fact, almost a century after Newton's publication, Young and Fresnel carried out some crucial experiments (for example, the double slit setup), and somehow combined Newton's ideas with Huygens' wave description. These works are considered by many to be the birth of modern Optics. It should be pointed out that this wave–corpuscle duality occurred in the theory of light, but it took another century before the works of Kirchhoff, Rayleigh, Jeans, and others on the emission and absorption of light led to the (apparent) dead end that P. Ehrenfest called the *ultraviolet catastrophe*, and which motivated a genuine revolution in Physics, where this problem of the nature of light resurfaced with strength [1].

The problem under consideration was the so-called black body radiation, that is, the study of the light emitted by a body heated to a temperature T. Physicists were interested in the distribution of energy and in the dependence of the radiated flux on temperature, since the experiments showed that the composition of the body was irrelevant, i.e., bodies with different compositions radiated in the same way if heated to the same T temperature. With the idea of calculating these quantities, physicists

© The Author(s), under exclusive license to Springer Nature Switzerland AG 2022   19
J. E. Horvath, *High-Energy Astrophysics*, Undergraduate Lecture Notes in Physics,
https://doi.org/10.1007/978-3-030-92159-0_2

considered the energy density of a radiant body, $\varepsilon(\omega)$, as a function of the frequency ω of radiated light (corresponding to the Huygens–Fresnel waves). Analysis of the classical problem of the frequencies of the waves inside a cavity that contains the radiant body indicates that the energy should reach what is known as equipartition, that is, a situation in which each frequency or mode has an energy $E = k_B T/2$, where k_B is the Boltzmann constant. Thus, multiplying this energy by the density of modes between ω and $\omega + d\omega$, we would have

$$\varepsilon(\omega) \propto \omega^2 k_B T d\omega , \qquad (2.1)$$

implying that the energy would grow without limit for high frequencies, which is physically impossible because the amount of energy radiated cannot be infinite. This inconsistency (or "catastrophe") pointed to some error in the basic hypotheses that needed to be clarified and corrected.

The interesting twist here is that the decisive idea to find a physical solution had already been expressed by Max Planck. For very different reasons, Planck had considered that the absorption and emission of radiation would happen discretely, in "packages" (or *quanta*) that satisfy the following relationship between energy and frequency [2]:

$$E = h\nu . \qquad (2.2)$$

When applied to the black body problem, this expression leads to a distribution that does *not* diverge for large ω. In terms of the frequency $\nu = \omega/2\pi$, this distribution is

$$B(\nu, T) = \frac{8\pi \nu^2}{c^3} \frac{h\nu}{\exp\left(\dfrac{h\nu}{k_B T - 1}\right)} , \qquad (2.3)$$

which is very different from the problematic "classical" form $(8\pi \nu^2/c^3)k_B T$.

The solution that avoids this "catastrophe" was thus implicit in Planck's hypothesis, even before it was calculated. Planck had worked on the notion of discrete packages without really believing in its physical reality, and had never been convinced of their factual existence right up until his death. It was Albert Einstein himself who elevated Planck's hypothesis of discrete quanta to the category of real physical objects, and hence very different from a mere mathematical trick [2]. Einstein was thus the "father" of quantum theory (although he never liked the consequences of the probabilistic interpretation that Bohr, Born, and others later developed) giving rise to the quantum of light, the photon, which satisfies (2.2) The application of this idea to another important problem, the photoelectric effect, not only confirmed the relevance of this approach, but also guaranteed Einstein the Nobel Prize in Physics of 1921. We will discuss the photoelectric effect and other processes involving photons in order to understand the instrumental developments that led to the growth of high energy Astrophysics in the following [3].

2.2   Processes Involving Photons at High Energies (Absorption and Scattering)

With the discovery of the photon or quantum of light, our perspective of the interaction of light with matter has undergone an important change. Although many phenomena were still well described with the wave formalism, the most elementary processes were better thought of as interactions between two particles, e.g., an electron, proton, etc., and a photon. This is not to say that light has ceased to be a wave phenomenon, nor that the idea of the quantum of light led to contradictions with previous results, but that in the microscopic world assigning a wave or particle character to an elementary quantum results in an inadequate and even confusing picture. We do not have any direct experience of the quantum world, and we have a (human) tendency to imagine that quantum objects should behave like something we do know from the macroscopic world, namely, waves and corpuscles.

With this quantum perspective, the study of absorption and scattering processes that involve photons interacting with matter has produced many important results, of direct application to the understanding of astrophysical processes and sources as we will see later. We now review some of these developments from the first half of the 20th century, then discuss the processes in which photons are emitted.

2.2.1   Photoelectric Effect

The young Albert Einstein carried out his research alone for several years while working in the Bern Patent Office, and besides coming up with the celebrated Special Theory of Relativity, he produced a simple explanation of the photoelectric effect using the bold hypothesis of the quantum of light. The basic observation that led Einstein to the latter was that electrons are ejected when a metal plate is illuminated with monochromatic light (a result that was already familiar to physicists at the time) and that there was a maximum speed of ejection that depended on the metal (Fig. 2.1)

What Einstein did next was to take Planck's photon hypothesis, which he had adopted as the true physical description of light, to its ultimate conclusion. In fact, Einstein's analysis began by admitting that there is a minimum amount of energy that must be delivered to pull an electron from the metal. This he called the work function W, which differs for each metal and results mainly from the action of the electrostatic forces that keep the electrons attached to it. The light that falls on the plate is assumed to be made up of photons of quantized energy $h\nu$. Thus, the energy available to accelerate the ejected electron is

$$E = h\nu - W . \tag{2.4}$$

The energy of the emitted electron is then measured using a voltmeter that registers a maximum voltage V_{max}, whence a maximum energy eV_{max} is observed, as already

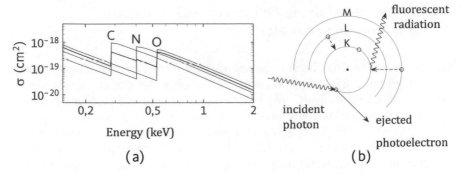

Fig. 2.1 **a** Photoabsorption cross-section for various materials as a function of energy. **b** Diagram showing the emerging fluorescent radiation when the incident photon energy is high enough

pointed out. To check this physical picture, it is enough to vary the frequency of the incident light and measure the $V_{\max}$ values for each metal that has a fixed W, that is,

$$e V_{\max} = h\nu - W . \tag{2.5}$$

A simple graph of the relation (2.5) made with the directly measured values of these physical quantities can be used to determine h/e and thus establish the very "core" of the quantum hypothesis as a verifiable corollary. Not only is the linearity of the relation shown to be accurate, but the numerical value of the Planck constant can be determined. In fact, Robert Millikan measured the latter with an accuracy of around 0.5% after independently determining the charge of the electron, and later received the Nobel Prize in Physics in 1923 for that work [4].

We can now explain the relevance of the photoelectric effect to high energy Astrophysics. As we will see below, if the photon energy is very high, other processes will be important, but for energies close to the typical value of the work function, photoelectric absorption can be used efficiently for shielding effects (for example, to protect astronauts from radiation). The material used must be heavy, since the optical depth, a measure of the probability of the photon interacting with the material and being absorbed, is given by

$$\tau = \int \sigma N_A dl \propto \frac{Z^4}{(h\nu)^3} , \tag{2.6}$$

where the cross-section σ is that of Fig. 2.1 (which shows the "jumps" due to the electronic layers), N_A is the number density of material A, and dl the differential distance along the line of sight. It is clear that for shielding, for example, against gamma rays ($E \geq 100$ keV), it is best to use material with a high Z value, such as lead.

In real processes, when the energy E is very high, photoabsorption is followed by *fluorescence*, since the electrons thrown out by the photon are from the innermost layers (for example, from the K layers), and soon other electrons de-excite, occupying the state of the one that was ejected and emitting the energy difference, also in the form of photons.

In Astrophysics, photoabsorption is often referred to as a *bound–free process*, with reference to the initial and final states of the ejected electron, regardless of the subsequent fluorescent emission that may or may not be present.

2.2.2 Compton Scattering

The study of the properties of light described in Chap. 1 saw important development in the early 20th century with the study of X-rays discovered by Röntgen in 1895. Their name already indicated a total ignorance of their true nature, but it soon became clear that they were actually highly energetic photons. The work of Barkla, Von Laue, and Bragg had shown that the X-rays are scattered in matter, but contrary to the prediction of classical Electromagnetism, their frequency also changes (besides their being deviated from the direction of the original beam). In 1923, Compton published a study in which he attributed a momentum to the quanta of light (photons) as if they were material particles, in total harmony with Einstein's initial ideas. Thus, the collision of a photon with an electron, impossible in classical theory, gained reality in the new Quantum Pysics initiated by Max Planck [4]. In Compton's work, the hypothesis of the momentum of the photon was consistent with (but not inspired by) the ideas of Einstein and Planck, and he proceeded to demonstrate that the frequency of the initial radiation would have to change when it collided with the electrons of a gas, using only the conservation of the momentum and the energy in the collision, as schematized in Fig. 2.2.

Since the Compton process must satisfy conservation of energy and of the two components of the momentum of Fig. 2.2, we obtain the following three conditions:

$$m_e c^2 + h\nu = E' + h\nu' , \tag{2.7}$$

$$\frac{h\nu}{c} = \frac{h\nu'}{c} \cos\theta + m_e \gamma \mathbf{v} \cos\phi , \tag{2.8}$$

$$\frac{h\nu'}{c} \sin\theta - m_e \gamma \mathbf{v} \sin\phi = 0 . \tag{2.9}$$

This is a system of algebraic equations that can be solved to find ν, ν', θ, and ϕ. After some algebraic manipulations and setting $\lambda = c/\nu$, we have

$$\frac{c}{h} = \frac{1}{\nu'} - \frac{1}{\nu} \longrightarrow \lambda - \lambda' = 2\lambda_C \sin^2 \frac{\theta}{2} , \tag{2.10}$$

where the quantity $\lambda_C = h/m_e c$ is the *Compton wavelength* of the electron. Numerically, $\lambda_C \approx 2.4 \times 10^{-4}$ Å is a very small number, and so the change in the frequency of light is also small. However, it is clear that $\lambda - \lambda' > 0$ and the photons lose

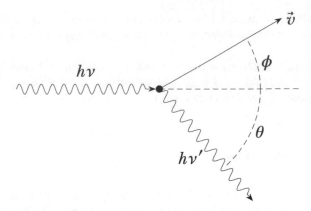

Fig. 2.2 Basic diagram of the Compton effect. A photon of initial energy $h\nu$ interacts with an electron at rest (in its own reference system) and is deflected through an angle θ from the initial direction, besides changing its energy to $h\nu'$. The electron acquires a certain velocity, with direction characterized by another angle ϕ. The problem here is to calculate the change in wavelength of the photon as a function of the angle, in order to compare with experimentally observed results

energy which was transferred to the electrons in the scattering process, thus known as *inelastic scattering*.

This process can be considered in the low ($k_\mathrm{B} T \ll m_e c^2$) or high ($k_\mathrm{B} T \gg m_e c^2$) energy limits. In the first, the cross-section cannot be sensitive to the energy of the photon, since it only "sees" the electron as a target with area of order r_e^2, where r_e is the *classical electron radius*, related to the Compton wavelength by the fine structure constant $r_e = \alpha \lambda_\mathrm{C}$. The cross-section at the low energy limit (called the *Thompson limit*) should be $\sigma \propto r_e^2$. The precise expression is

$$\sigma_\mathrm{T} = \frac{8\pi}{3} \left(\frac{e^2}{m_e c^2} \right)^2 , \tag{2.11}$$

as claimed, apart from the numerical pre-factor $8\pi/3$. In the opposite, high energy limit $k_\mathrm{B} T \gg m_e c^2$, the process depends on the incident energy and must be calculated using Quantum Electrodynamics. This goes far beyond the scope of our discussion, but we can quote the final result, the so-called *Klein–Nishina cross-section* (ultra-relativistic limit), which decreases with energy and is asymptotically smaller than σ_T (Fig. 2.3).

An important property of (2.11) is its frequency independence. As a consequence, within a certain range of energies, the Compton (Thompson) scattering will be the same for any incident photon.

In Astrophysics, we often have to deal with the *inverse Compton effect*, where an ultra-relativistic electron collides with a low-energy photon. This case has analogous mathematical expressions, and the desired result can be obtained by a simple Lorentz transformation between reference frames. As an example of this situation, Fig. 2.4

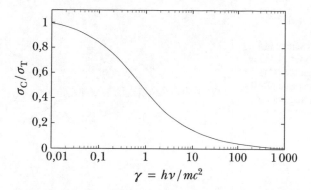

Fig. 2.3 Behavior of the cross-section in the Klein–Nishina limit. The decrease is clear for $k_B T \gg m_e c^2$ (right-hand region on the axis)

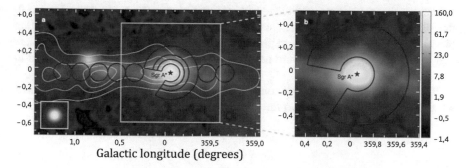

Fig. 2.4 Diffuse gamma emission from the center of the Milky Way for energies $E > 100$ MeV [5]. The measured spectrum requires the injection of protons of energy at least 10^{15} eV, possibly by the supermassive black hole Sgr A^* in periods of past activity. A fraction of the emission is not associated with the inverse Compton effect, but the latter is highly dominant. Credit: HESS Collaboration

shows the diffuse emission from the center of our galaxy attributed to the inverse Compton effect by ultra-relativistic protons accelerated and injected by compact sources in the region of the central source Sgr A^*.

2.2.3 Pair Production

As discussed in Chap. 1, the formulation of Quantum Physics significantly changed our notion of the vacuum and established that this is the state of minimum energy, but not really a "vacuum" in any classical sense, since there are permanent fluctuations and an energy density is attributable to it. A relationship between energy and mass (probably the most famous formula in the world) was obtained by Einstein in his Special Theory of Relativity:

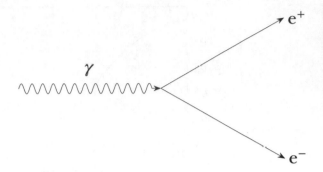

Fig. 2.5 When a photon
transforms into a
particle–antiparticle pair, it
must do so without violating
any conservation law. Can
we calculate the probability
of this process, and with it
the cross-section?

$$E = mc^2 , \tag{2.12}$$

a relation which, together with the idea of antiparticles in the quantum realm, led to
the following concept. Consider a (real) photon with high enough energy E. Equation
(2.12) would allow this photon to give rise spontaneously to a particle–antiparticle
pair by converting its energy into the pair's mass. Such a process is illustrated in
Fig. 2.5.

If ω is the frequency of the photon and $\gamma = 1/\sqrt{1 - (v/c)^2}$ is the Lorentz factor,
we can immediately write down the conservation of energy and momentum at the
vertex:

$$\hbar\omega = 2\gamma m_e c^2 , \tag{2.13}$$

$$2\gamma m_e v = \frac{\hbar\omega}{c}\frac{v}{c} . \tag{2.14}$$

As the initial momentum of the photon is $\hbar\omega/c$, and (obviously) $v < c$, it is impossible
to satisfy both conditions at once. Thus, the conversion of a photon into a pair *cannot
happen*. In order to satisfy the conservation laws what is needed is (1) a second
photon annihilating itself with the first one in the initial state (which is possible but
requires a radiation field with an enormous density) or (2) another agent that plays
the role of absorbing the additional momentum (usually a nucleus, Fig. 2.6).

In the presence of a Z-charged nucleus, we can consider the limit of low energies
by comparing the photon energy (in units of the energy of an electron with zero
momentum) with a dimensionless quantity related to the Coulomb energy. The "low"
energies are those that satisfy the condition $\hbar\omega/m_e c^2 \ll 1/\alpha Z^{1/3}$, or physically those
where the photon has enough energy to create the pair, but not enough to trigger more
complex effects related to the Coulomb field.

Under these conditions, an elaborate calculation using Quantum Electrodynamics
shows that the cross-section for pair production is

$$\sigma_{\text{pair}} = \alpha r_e^2 Z^2 \left[\ln\left(\frac{2\hbar\omega}{m_e c^2}\right) - \frac{218}{27} \right] \times 10^4 \text{ cm}^2 . \tag{2.15}$$

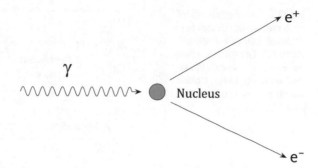

Fig. 2.6 Production of electron–positron pairs made possible by the presence of a nucleus that allows conservation of total energy and momentum

The things to note in this cross-section (2.15) are the quadratic dependence on the charge of the nucleus Z and the monotonic growth with the energy [3]. We have already seen that the Compton effect does not depend on the energy in the low energy limit and that the cross-section decreases at very high energies. Thus, it is inevitable that there should be a threshold above which pair production dominates the total cross-section of the photon interactions with matter.

In the limit of high energies $\hbar\omega/m_e c^2 \ll 1/\alpha Z^{1/3}$, the cross-section has a similar expression, with a weak dependence on the charge Z. This regime is hardly relevant and we shall not show the corresponding expressions.

2.2.4 Total Cross-Section and Absorption Coefficient per Unit Mass

Given the three processes discussed above, we can visualize the total result by introducing the *linear attenuation coefficient* as follows. We consider a beam initially with N_0 photons crossing a material of density ρ. There is a certain probability of interaction with the matter, which means that on average N of them will survive after traveling a distance x, where

$$N = N_0 e^{-\mu x} , \qquad (2.16)$$

and the quantity μ with dimensions of (length)$^{-1}$ is a linear attenuation coefficient that depends on the energy and the material. Multiplying and dividing by ρ in the argument of the exponential, we define μ/ρ as the attenuation coefficient per unit mass, generally expressed in $cm^2 g^{-1}$. If all the processes contribute to absorbing photons, the total attenuation coefficient results from the sum of each of them.

Figure 2.7 shows the regions where each of these processes dominates. (There is also coherent Rayleigh scattering, but it is never important in high energy Astrophysics, although it is responsible for the blue color of Earth's sky.) We see that the lighter the element, the broader the energies at which the Compton effect is dominant. Pair creation and the photoelectric effect will be important only at high and low energies.

Fig. 2.7 The regions where each process dominates the others as a function of the charge Z. The *lower dashed horizontal line* corresponds to magnesium (Mg), while lead (Pb) is near the top (see [6])

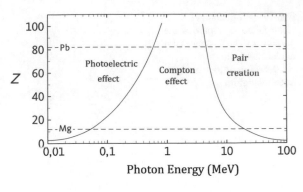

Fig. 2.8 Attenuation coefficient per unit mass for lead (Pb). Note the correspondence with the regions in Fig. 2.7. The *full curve* is the total coefficient dominated in each region by the processes indicated with the *dashed curves* (see [7])

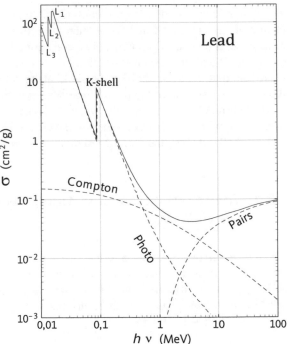

As a result of the different energy dependencies, the curves for the attenuation coefficient per unit mass behave as shown in Fig. 2.8. Although calculated for Pb, the shape is similar for any other element or compound substance. Thus, we can discuss the construction of photon detectors, whose interactions with matter we now know, and choose materials and configurations according to the range of energies we wish to observe. Essentially all these detectors need to be operated in space, since attenuation by the atmosphere, described by the same formalism, is otherwise inevitable. Note that all instruments already built and operated have used one of the forms of absorption discussed above.

2.3 Relevant Processes for High-Energy Photons (Emission)

While we have been discussing the fate of photons that find matter in their path, it is equally important to worry about the *emission* of photons by matter. In fact, this is precisely what we observe from high energy astrophysical sources. Thus, knowing the mechanisms of radiation emission is equivalent to obtaining a diagnosis of the physical conditions of the environment in which it originates.

Basically, the emission processes are classified as *coherent* or *incoherent*. The former have a very particular physical characterization: the particles emit collectively and the final amplitude reflects this collective character; usually it is proportional to the square of the number of emitters N^2. Incoherent processes, on the other hand, sum up random amplitudes of particles that emit individually, without correlation with their neighbors. We can study different astronomical sources by understanding the types of emission they produce.

2.3.1 Black Body Radiation

Black body radiation is the most basic example of incoherent emission. The term "black body" is the name used for a perfect emitter (and absorber), regardless of the actual "color" presented [4]. In fact, we have already considered this concept when discussing the origin of the quantization of light (photon). The spectral distribution per unit frequency is given by (2.3). An inspection of this expression (or in fact, a simple calculation) shows that there is a maximum of the function $B(\nu, T)$ for a certain frequency proportional to $k_B T$, and that the wavelength where the maximum occurs for a given temperature satisfies

$$\lambda_{\max} \times T = \text{constant} . \tag{2.17}$$

This is known as Wein's displacement law. This expression shows that the wavelength of the dominant radiation (i.e., the one most present in the incoming light) is inversely proportional to the physical temperature of the body. As the wavelength and frequency are inversely proportional, astrophysicists say that the body is "bluer" when its temperature increases, since the emitted radiation is typically more energetic (or "harder", because the photons are more energetic).

One of the most important characteristics of black body radiation is that the emission is independent of the composition, and is proportional to the fourth power of the temperature. In other words, the energy per unit area and time (flux F) is given by

$$F = \sigma T^4 , \tag{2.18}$$

where the proportionality constant $\sigma = 5.67 \times 10^{-5}$ erg cm^{-2} s^{-1} K^{-4} is called the Stefan–Boltzmann constant. As the emission is isotropic, the power, often called the *luminosity* L, is obtained simply by multiplying by the spherical area surrounding the emitting source, i.e.,

$$L = 4\pi R^2 \sigma T^4 . \tag{2.19}$$

Note that we will use the name "luminosity" instead of "power," which is the most common term in Physics courses, because the two are exactly the same and the former is standard in Astrophysics.

In most real situations, we will only have an idea of the luminosity of a source if there is a reliable estimate of its distance, since the flux is directly measurable by collecting the radiation that arrives from the object and identifying where the maximum emission is, as already discussed. As it is never possible to measure across the whole spectrum, we will almost certainly need to estimate how much we are leaving out, especially in the high energy bands, where a large part of the radiation can flow, outside the observed spectral range of our instruments.

2.3.2 Bremsstrahlung Radiation

The original German term *bremsstrahlung* can be translated as "radiation due to charge braking," but it is one of those relics of 20th century Physics that physicists and astrophysicists have held onto, and so we will continue to use it according to this tradition. From Maxwell's equations it follows that there will be electromagnetic radiation whenever an electric charge is subjected to acceleration, either positive or negative. It is this acceleration and the resulting emission that we will deal with here.

The most abundant and easy to accelerate or decelerate charges are undoubtedly electrons. The encounter of an electron with an intense electromagnetic field (typically the electrostatic field of a nucleus) deflects it from its original trajectory with resulting photon emission. Let us consider such a collision with impact parameter b, i.e., the prolongation of the approach trajectory would pass at distance b from the center of the target if it were not deflected. Applying Newton's law, the deceleration by the Coulomb field of the nucleus, assumed stationary due to its large mass, is

$$\mathbf{F} = m\mathbf{a} \approx -\frac{Ze^2}{x^2} , \tag{2.20}$$

that is, the magnitude of the acceleration is $a \approx Ze^2/mx^2$. Although the actual collision lasts quite a long time, the deceleration of the electron is effective only when it is very close to the nucleus, i.e., for a time $\Delta t \sim 2b/v$. The emitted power will be significant only during this time, and we can disregard the rest of the total collision time. This emitted power can be calculated using the *Larmor formula*, viz.,

$$P = -\frac{dE}{dt} = \frac{2}{3}\frac{e^2}{c^3}a^2 . \tag{2.21}$$

Fig. 2.9 A pulse of short duration, i.e., well located in time (*left*), extends to a certain maximum frequency (*right*). This maximum frequency can be estimated as $\nu_{max} \sim \Delta t/2 = v/4b$

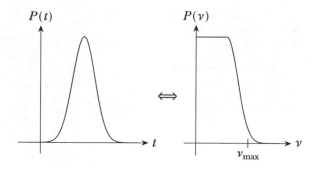

In the case of *bremsstrahlung*, by inserting the calculated acceleration into (2.21), we have

$$P = -\frac{dE}{dt} = \frac{2}{3}\frac{e^2}{c^3}\left(\frac{Ze^2}{mx^2}\right)^2 .$$

Taking into account the effective duration already indicated, we see that the emitted radiation consists of a "pulse" with energy

$$P\Delta t = \frac{2}{3}\frac{e^2}{c^3}\left(\frac{Ze^2}{mx^2}\right)^2 \times \frac{2b}{v} = \frac{4}{3}\frac{Z^2}{c^3}\frac{e^6}{m^2}\frac{1}{b^3v} . \tag{2.22}$$

The more precisely a pulse is located in time, the broader the frequency distribution (a result due to Fourier). The power emitted above a certain limit frequency is determined by the reciprocal of the pulse duration. The situation is illustrated in Fig. 2.9.

This value ν_{max} corresponds to the maximum energy of the colliding electron. If we divide (2.22) by the frequency interval $\Delta\nu$, we obtain the energy radiated per frequency interval in each collision:

$$\frac{P\Delta t}{\Delta\nu} \approx \frac{P\Delta t}{\nu_{max}} = \frac{16}{3}\frac{Z^2}{c^3}\frac{e^6}{m^2}\frac{1}{b^2v^2} . \tag{2.23}$$

These results refer to a single electron, but in realistic situations we must consider a set of electrons and ions with a distribution of initial energies. Let us consider the case of a "cloud" composed of electrons and ions (the latter all with charge Z for simplicity). Between two impact parameters b and $b + db$ there will be a number of ions equal to $2\pi n_Z v b db$, where n_Z is the number density of the ions (Fig. 2.10)

As the collision processes are totally independent from each other, the differential of the total number of collisions can be obtained by multiplying the previous expression by the number density of the electrons n_e, i.e., $2\pi n_e n_Z v b db$. Thus, we can integrate over all "rings" of radius b and obtain the *total emissivity*

Fig. 2.10 A cloud of
electrons that collide with
ions of charge Z. The
number of electrons and the
total number of collisions are
evaluated by the same
reasoning as for similar
problems in basic Physics

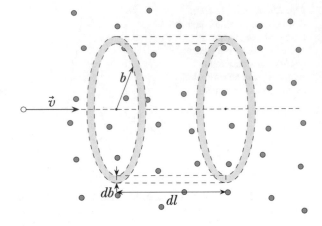

$$I = 2\pi n_e n_Z v \int_{b_{\min}}^{b_{\max}} \frac{16}{3} \frac{Z^2}{c^3} \frac{e^6}{m^2} \frac{1}{b^2 v^2} b db$$

$$= \frac{16}{3} \frac{\pi Z^2 e^6 n_e n_Z}{m^2 c^3 v} \int_{b_{\min}}^{b_{\max}} \frac{db}{b}$$

$$ZAS = \frac{16}{3} \frac{\pi Z^2 e^6 n_e n_Z}{m^2 c^3 v} \ln \left(\frac{b_{\max}}{b_{\min}} \right). \tag{2.24}$$

Note also that this result applies to a single particle velocity v, so we should integrate over v to include all the velocities present in the distribution. But before doing so, we will define the limits of integration of the expression (2.24) using physical considerations [3].

The value of $b_{\max}$ stems from the fact that the relevant interactions have impact parameters corresponding to $v < v_{\max}$ (Fig. 2.9 right). We thus have $b_{\max} < v/4v$. On the other hand, the minimum value of the impact parameter depends on the nature of the collision, since it may correspond to the classical domain or even be determined by Quantum Physics. In the first case, the validity of Newton's law is guaranteed by the condition $(Ze^2/mb^2)(2b/v) < v$, which implies $b_{\min C} \geq 2Ze^2/mv^2$, where the subscript C reminds us that we are considering a classical scenario. But if the collision is highly energetic, the classical condition may not apply. In this case the maximum approximation will be given by the uncertainty relation $\Delta p \times \Delta x \geq \hbar$, where the uncertainty in the position must be identified as the order of the impact parameter b. Thus, there is a $b_{\min Q}$ given by $b_{\min Q} \geq \hbar/mv$. The presence of the Planck constant $\hbar$ is a reminder of this situation, since this last $b_{\min}$ would be zero if the quantum of action was zero.

In most real cases in Astrophysics, electrons in the cloud have a classical velocity distribution, where we know that the characteristic velocity depends on the square root of the temperature according to $v = \sqrt{3k_B T/m}$. In this case the expression of the quotient in the logarithm of (2.24) can be simplified, and results to a good approximation in $b_{max}/b_{min} \approx (137/Zc)\sqrt{3k_B T/m}$. This numerical factor is usually built into the definition of the *Gaunt factor*: $g_{ff} = \sqrt{(3/\pi)}\ln(b_{max}/b_{min})$, a quantity always close to unity in free–free processes (with subscript ff). But to obtain the functional dependence of the emission, it is not enough to consider the typical value of the speed of the electrons, since there are collisions of electrons in the extremes of the distribution that can contribute in an important way. We then need to consider the whole distribution, which in the classical case is simply

$$f(v)dv = 4\pi \left(\frac{m}{2\pi k_B T}\right)^{3/2} \exp\left(-\frac{mv^2}{k_B T}\right) v^2 dv . \qquad (2.25)$$

We see that electrons with energies much greater than the average $k_B T$ suffer an exponential suppression, that is, they are progressively less present in the distribution. This expression should replace n_e, implicitly assumed to be monoenergetic in (2.24), and it should be integrated to evaluate the contribution of all electrons. The result, after inserting the numerical values of the physical constants, etc., is

$$I = 6.8 \times 10^{38} T^{-1/2} \exp\left(-\frac{hv}{k_B T}\right) n_e n_Z Z^2 g_{ff}(v, T) \, \text{erg s}^{-1}\,\text{cm}^{-3}\,\text{Hz}^{-1} . \quad (2.26)$$

We see that in the final result the factor corresponding to the power of a single electron $(T^{-1/2})$ is multiplied by Planck's cutoff $\exp(-hv/k_B T)$, and also by the densities of electrons and ions (assumed spatially homogeneous).

An electron and ion plasma that emits *bremsstrahlung* radiation has a characteristic cooling time in which it loses its energy E precisely by emitting the radiation. This time is

$$\tau = \frac{E}{I} = \frac{6 \times 10^3}{n_e \bar{g}_{ff}} T^{1/2} \, \text{yr} . \qquad (2.27)$$

Emission characterized as *bremsstrahlung* in galaxy clusters, containing thousands of galaxies, can be used to study the gravitational potential well of these clusters, which is due to both visible and dark matter. The argument is quite simple: the determination of the temperature associated with the *bremsstrahlung* from the spectrum of (2.25) serves to determine the motion of hydrogen masses, since $k_B T \sim m_H v_H^2$. For $T \sim 10^8$ K = 10 keV, this gives a speed of $v_H \sim 1000$ km s^{-1}. Only a very deep potential well can retain such a gas. When estimating this potential well by estimating the amount of visible matter, we find that it is much smaller than the amount necessary to retain the gas, and therefore almost 90% of the matter that produces the gravitational potential must be "dark". The example of the Coma cluster is shown in Fig. 2.11.

Fig. 2.11 The cluster of galaxies in Coma. This association shows the emission of *bremsstrahlung* attributed to the energetic electrons in the X-ray band [8]. Measurement of the spectrum indicates a characteristic temperature of around 10 keV. The observed matter (galaxies) falls far short of what is needed to explain why the gas does not escape the system. Credit: ROSAT/MPE/S.L. Snowden

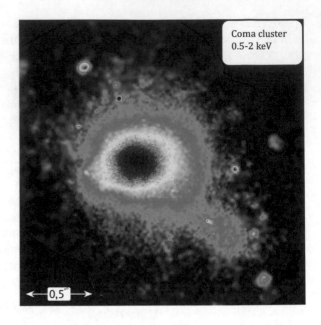

2.3.3 Synchrotron Radiation

Besides the radiation by the (de)acceleration of electrons in collisions, there is another way to produce radiation due to the presence of magnetic fields, which affect the trajectories of charged particles. The fundamentals of this phenomenon are contained in the famous Lorentz formula obtained, for example, in [4], which describes the force **F** on a charged particle q in the presence of electromagnetic fields **E** and **B**:

$$\mathbf{F} = q\mathbf{E} + q(\mathbf{v} \times \mathbf{B}) . \tag{2.28}$$

It is evident from the presence of the vector product in the second term that the component of the magnetic force is always perpendicular to the direction of **B**, and thus does not do work, but deviates the trajectory and thus contributes to accelerating the particle by changing its direction.

To better understand this behavior we can simplify the situation by ignoring the effect of the electric field, i.e., setting **E** = 0. This corresponds exactly to astrophysical situations, where it is almost impossible to generate a non-zero electric field (free charges would quench any such field almost instantaneously). We also assume that the magnetic field is uniform in the direction of the z axis (Fig. 2.12). For strong enough fields **B**, the particle paths are spiral, and the radiation produced at the expense of the particles in the **B** field emerges tangentially to the paths. This physical characteristic can be used to generate synchrotron radiation in the laboratory, by producing fields that confine electrons to move around a ring. Several experiments can then be mounted in tangential tunnels that make use of the emerging radiation (Fig. 2.12b).

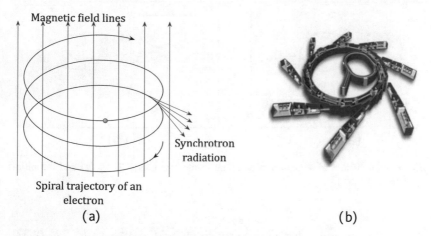

Fig. 2.12 Trajectories of an electron in a constant magnetic field in the z direction (**a**). Synchrotron accelerator (**b**). Note that the trajectories are circles as long as the momentum of the electron in the z direction is zero, which is guaranteed by the experimental setup. The tunnels contain targets of interest that are radiated by the synchrotron radiation discussed here. The world-class SIRIUS machine was inaugurated in Campinas (Brazil) in November 2018

We are interested in knowing the radiated power and the emerging radiation spectrum. We start by studying the motion of the charged particle in the **B** field, which satisfies the (relativistic) equation

$$\frac{d(m_0 \mathbf{v} \gamma)}{dt} = q\mathbf{E} + q(\mathbf{v} \times \mathbf{B}) . \tag{2.29}$$

This is analogous to Newton's equation, but with the presence of the Lorentz factor $\gamma = 1/\sqrt{1-(v/c)^2}$. Note that (2.29) is expressed in SI, not cgs units. As already pointed out, the particle has helical motion (Fig. 2.12a) with fixed pitch (launch) θ_0. The *cyclotron frequency*, which is the number of turns per unit time, is then

$$\omega_g = \frac{|q|B}{\gamma m_0} . \tag{2.30}$$

As the acceleration is purely perpendicular to the **B** field, we have $a_{\parallel} = 0$. The radiated power is obtained by applying the Larmor formula again, and yields

$$-\frac{dE}{dt} = \frac{\gamma^4 e^2}{6\pi \epsilon_0 c^3} |a_{\perp}^2| = \frac{\gamma^4 e^2 B^2 v^2}{6\pi \epsilon_0 m_e c^3} \sin^2 \theta_0 . \tag{2.31}$$

If we define the magnetic energy density $U_{\text{mag}} = B^2/2\mu_0$ and consider the non-relativistic limit $\gamma \to 1$, (2.31) becomes

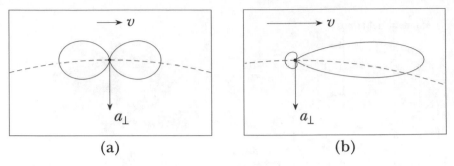

Fig. 2.13 Angular pattern of synchrotron radiation in the low energy limit (**a**) and the ultra-relativistic limit (**b**)

$$-\frac{dE}{dt} = 2\sigma_T c U_{mag} \frac{v^2}{c^2} \sin^2 \theta_0 = 2\frac{\sigma_T}{c} U_{mag} v_\perp^2 \ . \tag{2.32}$$

This is known as the *cyclotron power* (low energies). In the ultra-relativistic limit, on the other hand, the expression becomes

$$-\frac{dE}{dt} = 2\sigma_T \gamma^2 c U_{mag} \frac{v^2}{c^2} \sin^2 \theta_0 \ , \tag{2.33}$$

as a function of the pitch angle of the electrons. The angular pattern of the radiation is very much affected by the ultra-relativistic motion as compared with the low-energy limit (compare Figs. 2.13a,b).

Instead of considering an individual pitch, we can take an angular average over the momenta of the form $p(\theta_0)d\theta_0 = \sin\theta_0/2$, and write the power as

$$-\frac{dE}{dt} = \frac{4}{3}\sigma_T \gamma^2 c U_{mag} \frac{v^2}{c^2} \ . \tag{2.34}$$

The spectrum of each electron thus has a peak at $\omega = \omega_g$ and increasingly wide harmonics, until it fades and becomes a continuous (envelope). This situation is shown in Fig. 2.14.

It is now clear how to proceed to obtain the total emission from a population: if we have the electron density $n(E, r)$, the spectral density (total intensity per frequency interval) results from integrating spatially and in energy over all the contributions:

$$\frac{dI(\omega/\omega_g)}{d\omega} = \int_0^{E_{max}} \int_0^R -\frac{dE}{dt} n(E, r) \, dE \, dr \ . \tag{2.35}$$

An important and quite common case is that of an astrophysical source that injects electrons, accelerated by some mechanism, with a power law-type energy distribution, that is, $n(E, r) \propto E^{-\Gamma}$. Equation (2.35) can be immediately integrated to show that $I(\omega) \propto \omega^{-\alpha}$, with $\alpha = (1 - \Gamma)/2$. Observation of such a radiation distribution

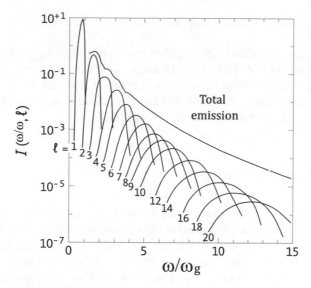

Fig. 2.14 Main peak (the maximum *on the left*) and harmonics of the synchrotron emission due to an electron. The *full line* shows the total emission

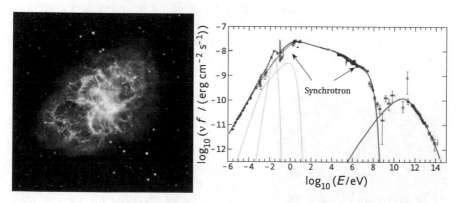

Fig. 2.15 Crab Nebula (*left*) and the spectral energy distribution (*right*), showing the regions where the synchrotron radiation is evidenced by its power law form [9]. Left credit: G. Dubner (IAFE, CONICET-University of Buenos Aires) et al.; NRAO/AUI/NSF; A. Loll et al.; T. Temim et al.; F. Seward et al.; Chandra/CXC; Spitzer/JPL-Caltech; XMM-Newton/ESA; and Hubble/STScI. Right: © AAS. Reproduced with permission

shows immediately that the population of injected electrons is not thermalized, since only a power law produces an $I(\omega)$ of this form. Hence, the source is *transparent* to the passage of electrons, which never interact enough to achieve a thermal distribution.

The best known case is probably the Crab Nebula (Fig. 2.15 left), where a young and energetic pulsar injects electrons into the environment, which in turn produces synchrotron emission when the latter move through the enormous magnetic field of the pulsar.

A closely related process is the so-called *curvature radiation*, in which the charge trajectories bend and then radiate for the same physical reasons (in fact, in the extreme relativistic limit the composit term *synchro-curvature* is found, in which both contributions occur together). We will not deal with curvature radiation here, and refer the reader to Longair's book for further discussion [3].

2.3.4 Čerenkov Radiation

In the first half of the 20th century, the Russian physicist P. Čerenkov studied a phenomenon that gave rise to the radiation that now bears his name. Čerenkov realized that, when a particle moves through a material medium (water, plastic, etc.), it can travel at a greater speed than light in that medium (but still less than c in vacuum). It is enough to remember that the simplest definition of the *refractive index* involves the quotient of the velocities of light in the medium in which it propagates. Thus, and in a manner completely analogous to the formation of a sonic shock wave (very common on the sea surface, for example), the resulting wave fronts at each point must combine to produce a *shock wave front*, and the energy transferred through it to the medium in the form of the excitation of the molecules produces electromagnetic radiation. The analogy is shown in Fig. 2.16.

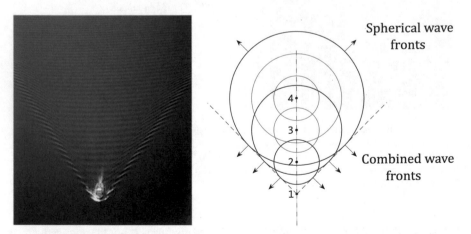

Fig. 2.16 Sonic analogue of Čerenkov radiation and its basic geometry. *Left*: Boat traveling at a speed greater than the speed of sound in water. The shock front formed by the combination of the spherical wave fronts along the path is clearly visible. *Right*: Charged particle (at the lower vertex) traveling at a speed greater than the speed of light in the medium. The combined wave fronts excite the molecules that radiate when de-excited and produce the Čerenkov light

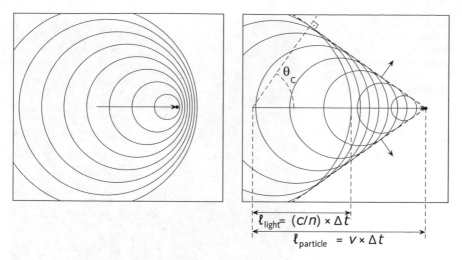

Fig. 2.17 Geometrical determination of the Čerenkov cone

To exemplify this phenomenon let us consider the passage of an electron through water. The dispersion relation (ratio of frequency to wave number) is

$$\omega = 2\pi \nu = \frac{2\pi (c/n)}{\lambda} = k\frac{c}{n} .$$ (2.36)

As the refractive index of water is $n \approx 1.3$, we see that the change in the dispersion relation from the one that in the vacuum, $\omega = kc$, allows us to infer the possibility of this kind of radiation.

The characterization of the so-called *Čerenkov cone* is simple and purely geometrical (Fig. 2.17). If the velocity of the charged particle is v, the distance traveled in an interval Δt is simply $l_{part} = v \times \Delta t$. In the same interval, the spherical front from the point where the particle was originally located has radius $l_{light} = (c/n) \times \Delta t$. Thus, the angle θ_C of the cone opening is

$$\cos \theta_C = \frac{(c/n) \times \Delta t}{v \times \Delta t} = \frac{1}{\beta n} .$$ (2.37)

We see that there is a physical requirement for this cone to form: as $\cos \theta_C$ must be a function with real values, the condition $\beta \geq 1/n$ must necessarily be satisfied. This is the kinematic condition (imposed on the speed of the particle) to observe Čerenkov radiation.

The intensity of the radiation depends on the excitation of the molecules, and therefore on the charge on the nuclei in the given medium, besides depending on the geometry itself, as determined above. Using these data, we can calculate the number of photons emitted per unit wavelength and per unit length along the path:

Fig. 2.18 Čerenkov radiation produced in a nuclear reactor. The medium (water) in which the charges move produces a bluish shine corresponding to (2.38)

$$\frac{\mathrm{d}^2 N}{\mathrm{d}\lambda \mathrm{d}x} = \frac{2\pi Z^2}{\lambda^2}\sin^2\theta_C \propto \frac{1}{\lambda^2}\,. \tag{2.38}$$

Thus, the largest number of photons will be in the region of smaller λ, and we therefore expect the color of the Čerenkov radiation to be in the visible blue band. This expectation is confirmed in the image of Fig. 2.18.

Čerenkov radiation is an important tool nowadays for the construction of detectors. For example, water tanks are used to measure the radiation produced by the passage of muons in cosmic ray showers (as we will see in Chap. 12). One can then reconstruct the energy and direction of arrival of the primaries.

2.3.5 Sources of Positrons and Pair Annihilation (Emission Lines)

To conclude this chapter, we now refer to the presence of radiation at definite frequencies, instead of spectral distributions, as was the case with each of the previous processes. The simplest and most frequent case is that of *electron–positron annihilation* lines, which is nothing other than the inverse of the pair production previously treated. For this to occur, some astrophysical source must eject the positrons, since the interstellar medium contains electrons in abundance.

What kind of mechanism can inject these positrons? A fairly common situation is that of an accelerator (Fig. 2.19) that can accelerate protons to high enough energy.

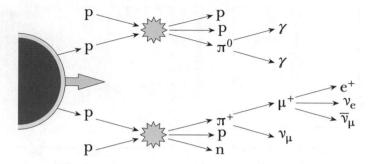

Fig. 2.19 An astrophysical accelerator injecting protons in the medium, producing positrons in the final state

The collision of these protons with surrounding material (nuclei) produces pions in the final state. The neutral pions π^0 give rise to gamma pairs (process shown in the upper part of Fig. 2.19), but the positively charged pions π^+ decay almost immediately into a positron and neutrinos. The latter positrons are the candidates to annihilate with electrons in the medium. In the case described, these positrons are actually "grandchildren" of the injected protons, so their number and the number of gammas must be correlated.

Figure 2.20 shows the electron–positron annihilation line observed from the direction of the center of the galaxy. The center of the line is practically at 511 keV, a value that corresponds to annihilation with a momentum of approximately zero. There is an interesting proposal for the origin of this line, besides injection by accelerators (still to be identified): the positrons could result from the decay of a particle that composes *dark matter*, if the latter happens to be unstable. There is no consensus or proof of this hypothesis so far, but it is certainly one of the most interesting problems that remains to be solved in high energy Astrophysics.

Another conspicuous and well identified source that produces positrons that subsequently produce the annihilation line is the decay of the aluminum isotope ^{26}Al. This isotope is produced in large amounts in explosive nucleosynthesis in supernovas. After $\approx 10^6$ yr, it decays sequentially into ^{26}Mg, producing a gamma-ray and a positron that annihilates. The presence of the annihilation line is well known in old SN remnants (see Chap. 6).

Electron–positron annihilation lines (and other lines in the spectra) do not always appear with the "standard" shape and in the expected position, such as the line in Fig. 2.20. There are several effects that can change the position and width of the line, the most important for our purposes being the *Doppler effect* and the *gravitational redshift*. The Doppler effect produces a broadening of the line and can produce a shift in its "natural" position if the emitter is moving in the direction of the line of sight. In the non-relativistic limit, the frequency difference $\Delta \nu$ satisfies $\Delta \nu / \nu = v/c$, and can shift the line toward the red or the blue, depending on whether the emitting material moves away or toward the observer, respectively. In the relativistic case the expression is more complicated and there is also a *transverse Doppler effect*, in

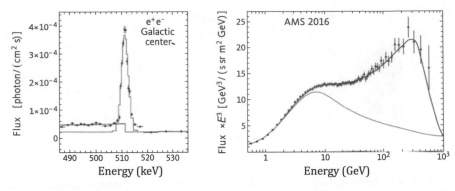

Fig. 2.20 *Left*: Annihilation line at the center of the galaxy. *Right*: The same phenomenon in the AMS-02 experiment [10]. The positron flow (*red*) far exceeds the expected proton production in the interstellar medium (*green*). Are there proton accelerators in this region or is it evidence of the decay of dark matter in the galactic bulge? A viable model is shown on the right with a *full black line*. Credit: AMS-02/CERN

Fig. 2.21 Gravitational redshift. The emission near the surface on the right of the object loses energy in its escape path to the observer on the right. The colors and the representation of the frequency in the figure reflect this phenomenon

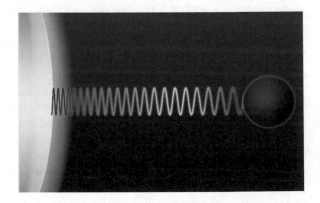

which the velocity component perpendicular to the line of sight also produces a shift in the frequency, although only a very small one.

Finally, gravitational redshift occurs when light escapes from a region where there is a very strong gravitational field. In fact, photons lose energy through the action of the field that "pulls" them in the central direction (Fig. 2.21). If emission occurs at the source with frequency v, the observer "at infinity" will detect a lower frequency

$$v_\infty = v\left(1 - \frac{GM}{Rc^2}\right)^{1/2}. \tag{2.39}$$

We see that measurements of frequency that differ from the "natural" value (for example, 511 keV in the case of electron–positron annihilation) can reveal characteristics of the object that produces the field (more specifically, the mass-to-radius ratio). It is important to clarify that this displacement is *not* related to the cosmological redshift, of a very different origin and interpretation.

References

1. S.F. Mason, *A History of the Sciences* (Collier Books, New York, 1973)
2. A. Pais, *Subtle Is the Lord: The Science and the Life of Albert Einstein* (Oxford University Press, Oxford, 2005)
3. M. Longair, *High-Energy Astrophysics* (Cambridge University Press, Cambridge, 2011)
4. R. Eisberg, R. Resnick, *Quantum Physics* (Wiley, Hoboken, NJ, 2004)
5. H.E.S.S. Collaboration, Acceleration of peta-electronvolt protons in the galactic centre. Nature **531**, 476 (2016)
6. W.R. Hendee, E.R. Ritenour, *Medical Imaging Physics* (Mosby-Year Book, St. Louis, 1992)
7. MIT OpenCourseWare, https://ocw.mit.edu/courses/nuclear-engineering/22-101-applied-nuclear-physics-fall-2006/
8. U.G. Briel et al., A mosaic of the Coma cluster of galaxies with XMM-Newton. Astron. Astrophys. **365**, L60 (2001)
9. F. Aharonian et al., The Crab Nebula and Pulsar between 500 GeV and 80 TeV: Observations with the HEGRA stereoscopic air Cerenkov telescopes. Astrophys. J. **614**, 897 (2004)
10. AMS02, https://ams02.space/de/node/474

Chapter 3
Detection and Instrumentation
in High-Energy Astrophysics

3.1 Spatial, Spectral, and Temporal Domains

To study sources in High Energy Astrophysics there are three different domains,
all of which reveal their nature and physics through observations. When we say
"domains" we are referring to the observed quantities, and more particularly, how
these depend on the variable that characterizes them (position, energy, or time). The
basic problems and techniques of X-ray and gamma-ray instumentation and their
solutions are briefly presented here.

The *temporal domain* consists in the study of the *variability* of some quantity
measured as a function of time. Typically, the photon count is measured in some
frequency range (for example, the whole band covered by a detector, or some more
specific channel of it), and the graph as a function of time reveals the temporal history
of the emission. By analogy with optical astronomy, this type of graph is called a
light curve (although the "light" here typically refers to X-rays or gamma rays). An
example of this type of observation is shown in Fig. 3.1 (right).

The light curve has the potential, among other possibilities, to reveal the physical
size of the actual emission zone. The reasoning is as follows: if τ_{min} is the smallest
time scale observed in the light curve, the region of emission is limited to a size
$R \leq c \times \tau_{min}$, otherwise we would be in the presence of an emission that violates
causality. We can say that, within a scale R, the smaller elements of the source can be
in causal contact to produce the emission, but not those outside this scale. Thus, for
example, a variability of, say 0.1 ms implies a maximum size for the emission region
of about 3 km. We see that this points directly to a compact object as a source of
radiation, although the whole object may not necessarily participate in this emission.
There may be, for example, just a "hot spot" emitting, or some other limited region
of the system.

Another important characteristic of sources is revealed by their spatial location
(*spatial domain*). This is particularly relevant when studying binary systems, since
it is important to know which component is emitting, whether it is the donor (sec-

© The Author(s), under exclusive license to Springer Nature Switzerland AG 2022   45
J. E. Horvath, *High-Energy Astrophysics*, Undergraduate Lecture Notes in Physics,
https://doi.org/10.1007/978-3-030-92159-0_3

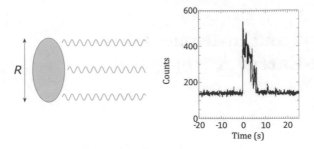

Fig. 3.1 Light curve of a variable object (in this case, a gamma-ray burst) *on the right*, and inference of the physical size of the source (*left*, see text)

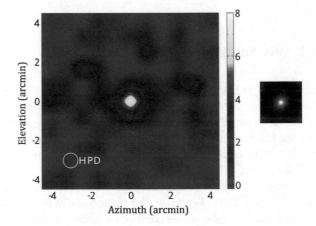

Fig. 3.2 *Left*: The source Cyg X-1, obtained with the HERO experiment on board a balloon with an angular resolution of a few minutes of arc (arcmin). Credit: NASA/Marshall Space Flight Center. *Right*: The same source observed with the Chandra satellite, reproduced on approximately the same scale, where the resolution reaches about 1 arcsec. Credit: NASA/Chandra X-ray Observatory

ondary), the receiver (primary), or the gas that passes from one to the other (accretion disk/shock), and for many other cases, in other real sources. The better the spatial resolution of the source, i.e., the ability to locate photons spatially, the better we can understand the geometry of the source (although for very compact or very distant sources this separation may be impossible). With some license, we could equate the spatial resolution to the "sharpness" of an optical image. In the following, we will discuss the difficulties of focusing when the photon energy is very high, but for now it is enough to point out that the first X-ray detectors only saw "spots" of various degrees, while the most modern observatories work with a resolution comparable to optical telescopes, but on X-rays (of order one second of arc, or arcsec). An example of the substantial improvement in the spatial resolution over time of X-ray observations is shown in Fig. 3.2.

The last domain is the *spectral domain*, represented graphically with the photon energy on the abscissa. This is used to study the way the source distributes the emitted energy, data which can be linked to physical characteristics through models. Here it is

Fig. 3.3 *Upper*: Spectrum of a coronal mass ejection passing by the Earth, obtained by XMM-Newton [1]. The spectrum shows highly excited elements (OVIII, MgXI, etc.) whose lines appear well resolved. Credit: Univ. Leicester, Dept. Physics and Astronomy. *Lower*: Example of the spectrum of a neutron star during a burst [3]. Several important emission lines are observed, contributing to our understanding of the nature of the source, even though their reality has been questioned. Credit: XMM-Newton/RGS

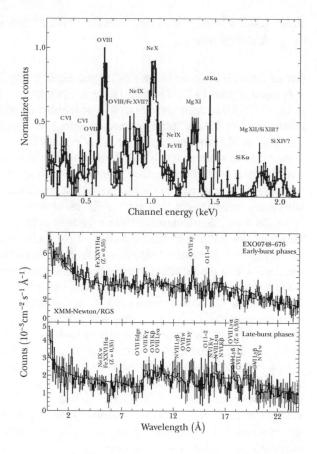

the spectral resolution ($R = E/\Delta E$) that matters. Hence, the most convenient thing is to be able to determine the energies of precisely detected photons. Any particularity in the spectrum (for example, the presence of spectral lines) can be used to make a diagnosis. In addition, one may be able to separate various emission components associated with different processes, perhaps originating in different regions of the source. Figure 3.3 shows two examples of resolved spectra from known sources.

Instruments are always built to detect a range of energies, with good time resolution (usually determined by on-board electronics), and an appropriate angular resolution, although the latter presents some difficulties that we will discuss below. Another characteristic of great importance is the collecting area of the instrument, since sources are often weak in the X and γ bands, and the limited numbers of photons arriving from them must be used as efficiently as possible. This makes it important to have large collecting areas, but taking into account compromises with other intrinsic characteristics of the instrument (see below).

3.2 CCDs in Optical Astronomy and High-Energy Astrophysics

One of the most important advances in optical astronomy, developed and popularized very rapidly since 1970, is the solid state detector known as a *charge-coupled device* or CCD [2]. The device consists of a large number of photon-sensitive zones called *pixels*, used to form a spatially accurate image of a region. It can also obtain the distribution of incoming photon energies and perform spectroscopic analysis that is essential to understanding the physics of the emission. Today most people are familiar with pixels because they actually own digital television sets and cameras. Although these commercial CCDs are much less reliable than the scientific ones and have many more defects, they work in the same way.

CCDs are made of a semiconductor (usually silicon), while the pixels are determined by the position of the electrodes above them, as indicated by I01, I02, and I03 in Fig. 3.4. A positive voltage is applied to the electrodes (as shown for I02 in the figure), and the resulting electric potential attracts electrons to the area below the electrode (little blue balls in the figure), while the positively charged holes are repelled (little red balls in the figure). Thus a *potential well* is generated where the electrons produced by the incoming photons accumulate. As more photons arrive, the well accumulates electrons until it fills up completely. It is important not to exceed this limit. The signal must be integrated (i.e., allowed to accumulate), but it must not exceed the capacity of the well, because the image to be generated would be distorted (astronomers speak of a *saturated image*). The most common type of CCD in astronomy has 1024×1024 pixels, although special configurations can be made. Taking into account the fact that an ordinary pixel is around 10–20 μm, the physical size is about 2 cm^2. Depending on the application, the pixels can be made much larger and associated in large mosaics.

Each CCD pixel is affected by three electrodes (see Fig. 3.4). We have already talked about the need to create the potential well, but the other two are needed to transfer the accumulated charge out of the device. For this, each electrode is kept at high and low voltage alternatively, to transfer the charge to the neighboring pixel in row or column mode depending on how the electrodes are oriented. For this reason (the transfer of charge from one pixel to the next and so on until the end), it is said that the charges are *coupled*, and hence the name CCD. The final reading of each pixel is taken by means of an amplifier that converts the accumulated charge into voltage,

Fig. 3.4 *Left*: A scientific grade CCD with a physical size of approximately 2 cm on the side. *Right*: Profile of the CCD structure

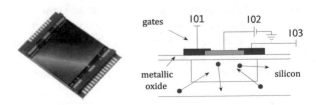

typically a few μV for each one. This way, even a voltage of a few volts requires the reading of about 100,000 units in each pixel. A CCD camera thus consists not only of the CCD, but also of the associated electronics which reads all the pixels, removes noise (electrons that have nothing to do with the source), and digitizes the signal (the CCD is in fact an analog device), and software to analyze the data and create images from them.

An important parameter in the detections is the so-called *quantum efficiency*, that is, how many photons are actually detected for every 100 that occur. The human eye is a very inefficient detector, capturing only around 20% of the incident photons. Photographic films are even worse, with efficiency around 10%. But CCDs can easily achieve efficiencies of 80% or more (depending on the measured wavelength). This characteristic, together with its mechanical robustness and simplicity of operation, justify its enormous and rapid acceptance in astronomy, even more so when considering sources that only emit a small number of photons that have to be exploited to the maximum.

Optical CCDs are generally sensitive to photons of the whole visible spectrum and some part of the infrared. But it is possible to manipulate the construction to extend the sensitivity to shorter wavelengths into UV and X-rays. It is precisely this feature that interests us.

To obtain complete data from an observed region it is important to know the range of source intensities (i.e., the faintest and brightest) that can be recorded. The detection threshold is usually placed some 3σ above the device noise (in the statistical sense of signal significance). As already pointed out, the maximum intensity is that which "fills" the potential well without saturating it, typically around 100 000 electrons. The observation range is defined, taking into account the fact that in the very process of reading the pixels an irreducible noise is generated (generally a small count of 3–4 electrons). The quotient of the maximum to the minimum counts is called the *dynamic range*.

The whole discussion so far has remained quite general, but our specific interest in the higher energies deserves special consideration. Optical CCDs have a very desirable feature for any type of electronics: they are linear devices, where one incident photon produces one electron. Thus, the reading is directly proportional to the number of incident photons. However, when we deal with X-rays, this is no longer true. The incident photons are much more energetic and each produces a multiplicity of electrons, in greater numbers the higher the energy (typically between 100 and 1000 electrons). CCDs should thus operate in the region of non-linear response, which is not a disaster, but requires a more sophisticated treatment to produce the final images, since the intensity is no longer directly proportional to the number of incident photons. Another difference is that, although the efficiency increases with pixel size, it is not possible to take advantage of this trend for X-rays, since the charge fatally begins to be deposited in more than one pixel, thus losing spectral resolution (even though methods can be applied to recover the maximum amount of information).

We see that the spectral resolution $R = E/\Delta E$ and also the temporal resolution Δt suffer from limitations when using CCDs. Additionally, the type of telescope is important for detection, as we shall now discuss.

3.3   The Problem of Focusing (Imaging) High-Energy Photons and Its Solutions

When dealing with optical refracting telescopes there are several ways to focus the light, two of which are shown in Fig. 3.5. In fact, Newtonian and Cassegrain foci are widely used, and due to the relative scale of the wavelength of the observed light, there are no technological problems that prevent an accurate focus.

However, in the treatment of X-ray telescopes this issue takes on much greater importance, since the wavelength is much shorter and focusing is more difficult. In other words, an image of any X-ray object will be "out of focus" unless we can build telescopes that solve this focusing problem. For these purposes, it was necessary to explore the basic physical properties of hard photons and build new designs that allow efficient imaging.

The first physical property required is *Snell's reflection law*. An incident ray from the vacuum with refractive index n_1 on top of a reflective material that has refractive index n_2 is subject to a geometric deviation in the transmission that satisfies

$$\sin \theta_T = \frac{n_1}{n_2} \sin \theta_i , \tag{3.1}$$

where θ_T and θ_i are the transmission and incidence angles, respectively. Since $n_1 > n_2$ because photons arrive from the vacuum onto the reflective material, there will be real values for transmission only in the case of angles $\theta_i < \arcsin(n_2/n_1)$. Choosing the reflective material, e.g., gold with $n_2 = 0.99$, there will only be transmission for angles $\theta_i < 81.9°$. If incidence occurs at a higher angle, Snell's law will not be satisfied and there will be no transmission (see Fig. 3.6).

This feature is exploited by building a *grazing incidence* reflector arrangement, hereafter denoted by GI (Fig. 3.7). In each set of reflectors, light from the source is deflected until it can focus on the detector [5]. In fact, there are several complications in this design that we will not address here. The important thing is that this shows how the X-rays can be focused.

Fig. 3.5 Two common configurations of optical telescopes: Newtonian (*left*) and Cassegrain (*right*) [4]

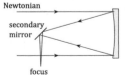

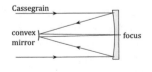

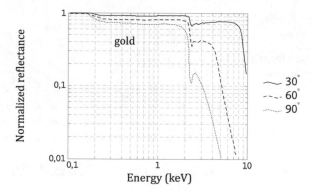

Fig. 3.6 Reflectance as a function of energy (*horizontal axis*) for three different values of the incidence angle

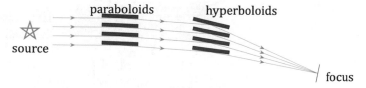

Fig. 3.7 Basic scheme for a design using a GI setup for X-ray telescopes

There is second popular method for focusing X-rays, using a more conventional incidence arrangement. This consists in stacking 50–500 alternating layers of platinum and carbon or tungsten and silicon, or any other combination of high-Z layers interspersed with low-Z layers. The aim is to take advantage of the constructive interference of the fronts that satisfy the Bragg condition

$$2d \sin \theta = n\lambda \,, \tag{3.2}$$

where d is the separation of the ions in the solid lattice, θ the angle of incidence, λ the wavelength of the radiation, and n an integer. The central idea is to get the radiation beams to acquire a phase difference of 2π and thus add up the intensities when reflected in the multi-layer "sandwich" (Fig. 3.8). As we have already said, the incidence need not be shallow, and with careful construction the reflectance can exceed 80%. This solution is referred to as a normal incidence (NI) setup.

Finally, when the energy of the incident photons increases still further, there will be no way to focus with these settings. Thus, photons with energies above 100 keV are detected using the *scintillation* technique. Scintillation (Fig. 3.9) consists in detecting the emission of secondary light as a result of the interaction of the high-energy photons from the sources when they are absorbed in a crystal or liquid. Photomultiplier cells convert the emitted light into a current that can be read and translated into the primary photon energy.

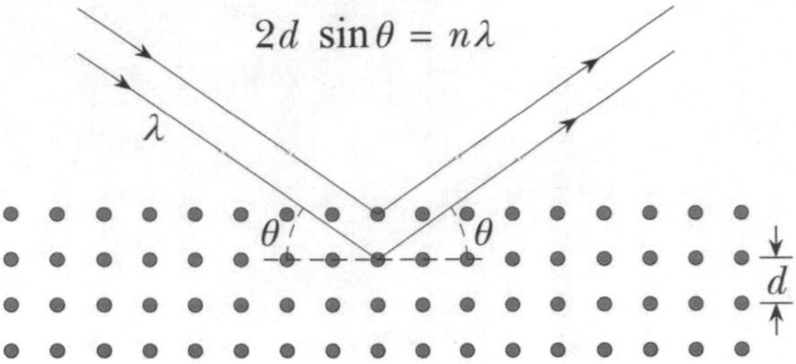

Fig. 3.8 Bragg diffraction for beams with normal incidence NI (i.e., non-grazing). When Bragg's condition is satisfied, the amplitudes add up in the detector

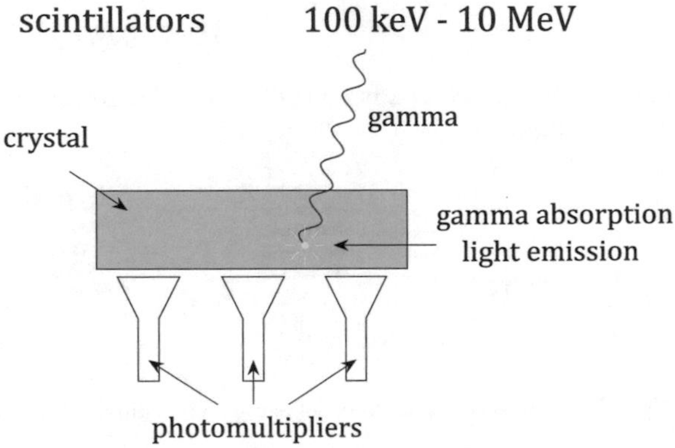

$$2d \sin\theta = n\lambda$$

Fig. 3.9 Photomultiplier detector

scintillators   100 keV - 10 MeV

crystal   gamma

gamma absorption
light emission

photomultipliers

This technique can be used to oberve photons of up to 20–30 MeV (see next section). But far above this value, in the TeV range or higher, one must work with the Čerenkov technique. We have already seen in Chap. 2 that the passage of a charged particle through a material medium excites the molecules in that medium, whereupon they decay and emit light. But a photon also produces a similar effect, although the spatial pattern of radiation is different and can be distinguished. Thus, the Čerenkov effect provides a tool for studying very high energy gamma sources such as AGNs (Chap. 8).

3.4 Space-Based Instruments for X-Ray and γ-Ray Detection

The beginning of X-ray Astrophysics (the "softest" photons in the high energy region) had to await the development of space technology. Shortly after the launch of Sputnik in 1957, a group of scientists led by Riccardo Giacconi launched an X-ray detector on board an Aerobee 150 rocket and discovered the first X-ray source outside the Solar System, called Sco X-1. The angular resolution was so poor (worse than 20°) that it took time to identify the "spot" in X-rays with the constellation of Scorpius. Today we know that it is a neutron star with a low mass companion (LMXB), the first

Table 3.1 Some of the most important X-ray astrophysics missions

Name	Type	Effective area (cm^2)	Resolution (arcsec)	Energy band (keV)	Observations
Einstein (1978–81)	GI	~ 200 at 1 keV	~ 15	0.2–4.5	Discovered ~ 7000 sources
ROSAT (1990–99)	GI	~ 400 at 1 keV	~ 5	0.1–2.4	Discovered ~ 150,000 sources
ASCA (1993–2001)	GI	~ 1300 at 1 keV, ~ 600 at 7 keV	~ 175	0.5–10	4 independent telescopes
BeppoSAX (1996–2002)	GI	~ 330 at 1 keV	60	0.1–10	
Chandra (1999–)	GI	~ 800 at 1 keV	0.5	0.1–10	Best spatial resolution
XMM (1999–)	GI	~ 4650 at 1 keV, ~ 1800 at 10 keV	14	0.1–12	4-ton mission
NuSTAR (2012–)	GI	~ 850 at 9 keV	< 10	3–78	

example showing how the sky is populated with high energy sources. A chronology of the most relevant missions can be found at the website [6].

A few years later in 1970, the first satellite dedicated to X-ray sky exploration was launched. This was the Uhuru mission (meaning "freedom" in Swahili), with an effective area of only 0.084 m^2 and coverage in the 2–20 keV band, capable of a spatial resolution of about 0.5°. Uhuru identified more than 300 sources, among them Cyg X-1, the first black hole candidate in our galaxy. Over time, several missions have been launched, some of them still in operation, and the exploration of X-ray sources has continued on a sustained basis. Table 3.1 shows some of the most important space missions in X-ray astrophysics, along with their most important features (collecting area, spatial resolution, and energy band).

As previously mentioned, one must use scintillators to study still higher energies. The Compton effect described above was the basis of the COMPTEL instrument, containing liquid scintillators above a NaI crystal. The successive interactions of the incident photons until they are finally absorbed can be used to determine the direction of their arrival without having to actually focus them in the conventional sense. The coded mask technique was developed to improve the spatial resolution, which is made much more difficult by the fact that gamma rays cannot be focused. This is an advanced variant of the *camera obscura* used in the Renaissance. It was

Fig. 3.10 Coded-mask
concept, used to image γ-ray
sources up to 20–30 MeV

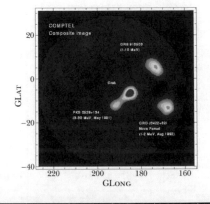

Fig. 3.11 Spatial resolution
in gamma rays and its
consequences. *Upper*:
COMPTEL image of the
Crab Nebula (the brightest
source), Geminga (top right),
and the extragalactic source
PKS 0528+134 (bottom
right), with a resolution of
around 1°. Courtesy of the
COMPTEL Collaboration.
Lower: Image of the galactic
center made by the IBIS
imager aboard the
INTEGRAL mission with a
much higher resolution of
about 12 arcmin [8]. Sources
are clearly separated at
angles where they would be
confused by the resolution of
COMPTEL. Credit: A.
Paizis (ISDC, Geneva &
IASF/CNR, Milano)

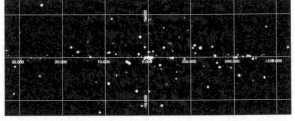

first proposed in the 1960s. The photons pass through a series of holes of known
pattern in a mask, thus forming a "shadow" in the plane of the scintillation detector
(Fig. 3.10). Algorithms process this "shadow" image, using the fact that its exact
geometry is known, and the image can be very substantially improved. This method
is used to image sources in gamma energies [7].

The most recent example of this type of instrument, still in operation, is the IBIS
imager in the INTEGRAL mission, which achieved an angular resolution of about
12 arcmin for energies of around 10 MeV. This makes a huge difference when it
comes to identifying sources individually, although an even higher resolution would
be desirable (Fig. 3.11).

Finally, for even higher energies, the technique must be changed once again to take
advantage of the Čerenkov effect, detecting the radiation produced by the passage of

Fig. 3.12   The MAGIC telescopes in the Canary Islands, built in 2004 to study cosmic rays at the highest energies [9]. Credit: Giovanni Ceribella, Max Planck Institute for Physics, Munich

particles through the atmosphere or water. The MAGIC telescopes shown in Fig. 3.12 exemplify the use of this type of instrumentation, operating between 25 GeV and 30 TeV, with a large collecting area (and the addition of a second, similar instrument for stereoscopic data acquisition). Essentially all energies of this level can be detected using the Čerenkov effect, and in Chap. 12 we will see how it can be combined with detectors of another type, in fact, water tanks, to study cosmic rays.

We have addressed the main features of photon radiation arriving from high-energy sources, but in the 21st century there are new probes of non-electromagnetic character that can reveal the high energy universe and its sources. In Chap. 9, we shall discuss neutrino astronomy and in Chap. 10 the recent field of gravitational waves. When combined with the "old" techniques of photon astronomy, these new signals provide a very powerful integrated view of sources and phenomena. It can be said that 21st century High-Energy Astrophysics is to a large extent a *multimessenger* discipline, going well beyond the traditional realm of photon astronomy.

References

1. J.A. Carter, S.F. Sembay, Identifying XMM-Newton observations affected by solar wind charge exchange—Part I. Astronomy and Astrophysics **489**(2), 837–848 (2008)
2. D.H. Lumb et al., Charge coupled devices (CCDs) in X-ray astronomy. Experim. Astron. **2**, 179 (1991)

3. J. Cottam, F. Paerels, M. Mendez, Gravitationally redshifted absorption lines in the X-ray burst spectra of a neutron star. Nature **420**, 51 (2002)
4. See, for instance, https://www.open.edu/openlearn/science-maths-technology/telescopes-and-spectrographs/-content-section-1.4
5. P. Murdin (ed.), *Encyclopedia of Astronomy and Astrophysics* (Nature Publishing Group, London, 2001)
6. https://heasarc.gsfc.nasa.gov/docs/heasarc/headates/heahistory.html
7. M.J. Cieślak, K.A.A. Gamage, R. Glover, Coded-aperture imaging systems: Past, present and future development. A review, Radiation Measurements **92**, 59 (2016)
8. Paizis et al., *Proceedings of the 5th INTEGRAL Workshop*, The INTEGRAL Universe (2004), https://sci.esa.int/web/integral/-/37398-ibis-isgri-observations-of-the-galactic-centre-region
9. J. Cortina, Highlights of the MAGIC Telescopes, in *32nd International Cosmic Ray Conference* (2011)

Chapter 4
Stellar Evolution up to the Final Stages

4.1 Stellar Astrophysics

The nature of the observed stars has been a subject of discussion and speculation since the early days of civilization. Atomists Leucippus and Democritus thought that the Milky Way was made of stars, which they considered too small to be distinguished from one another. By the time of the Indian mathematician Aryabhata (5th century A.D.), there existed in the East the notion that stars were, in fact, other suns. It would have been immediately obvious that they would have to be at enormous distances for this hypothesis to make sense. Other important speculations were formulated in the West. For example, in Giordano Bruno's writings, not only were the stars identified as distant suns, but inhabited planetary systems accompanied them, putting the author on a direct collision course with the Roman Catholic Church. What is certain is that it was only in the early 19th century, with the works of W. Herschel and J. von Fraunhofer, that the star = Sun identification was shown to be correct: the absorption lines of several nearby stars were observed, revealing their kinship with the lines observed in the solar spectrum. This Chapter addesses the construction of stellar models and the important features of Stellar Evolution till the final stages leading to explosions/compact object formation.

Although the nature of stars remained for many centuries on a speculative plane, scientists from Classical Antiquity devoted themselves to their study. The first catalog of stars created in the West was authored by the mathematician and astronomer Hipparchus, and contained some 850 stars observable by naked eye, as reproduced in Ptolemy's *Almagest*. Hipparchus and other later astronomers also noticed the differences in brightness of the stars, and especially in their colors (Fig. 4.1). These ancient observations and those recorded after the invention of the telescope in the early 17th century led directly to the basic questions of stellar Astrophysics that will be the subject of our discussion: Are stars "eternal"? What is their internal constitution? How can these questions be linked with available observations? The enormous development of the theory of Stellar Evolution throughout the 20th century

J. E. Horvath, *High-Energy Astrophysics*, Undergraduate Lecture Notes in Physics, https://doi.org/10.1007/978-3-030-92159-0_4

Fig. 4.1 Star field around
the constellation of Cygnus,
made up of 33 images
obtained at the Mount
Palomar Observatory using
two filters, showing a variety
of colors (real) and the
brightness of the stars in this
field. Credit: Palomar
Observatory/California
Institute of Technology

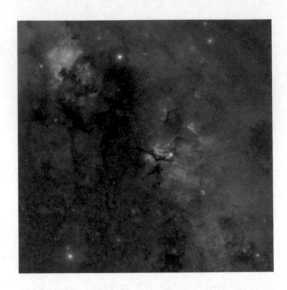

and the state of the art in this field will occupy the rest of this chapter and part of the
following chapters.

4.2 Basic Facts and Observations

As already pointed out, the identification of lines in stellar spectra showed that stars
were objects of the same type as the Sun. The presence of known chemical elements
(H, C, O, etc.) and the study of spectral lines gives information about the outer stellar
region, while its inner composition remains undetermined because the radiation does
not carry information from the inner regions. But there are other ways to study stellar
structure, at least indirectly. For example, it is relatively easy to determine how much
energy is flowing from a star per unit frequency interval. In order to do this, it
is enough to use filters that let through photons in some chosen range, and then
count how many of them are in each wavelength interval. The total energy is easily
calculated with the aid of the relation $E = h\nu$. In general, and for "normal" stars
(the case of white dwarfs and others will be treated later), we find that there is a
band where the number of photons reaches a maximum, and that the spectrum has
approximately the shape of a black body spectrum. The black body, discussed by
G. Kirchhoff and others at the beginning of the 20th century, is an idealization that
applies to a perfect absorber/emitter which presents the distribution of intensities as
a function of frequency shown in Fig. 4.2.

The colors of the stars correspond to the maximum of the distribution, since the
photons with this wavelength are the most numerous, provided that the star complies
with the black body idealization. The value of λ_{max} moves to lower wavelength

Fig. 4.2 Radiation intensity as a function of wavelength ($\lambda = 1/\nu$) for three "stars" (idealized as black bodies) whose effective temperature is indicated. This effective temperature corresponds to the wavelength where the maximum occurs

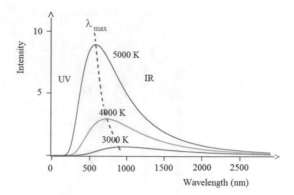

values as the temperature increases. The lower the value of λ_{max} (i.e., the higher the frequency ν_{max}), the higher the temperature. Thus, the photons are said to be "harder" (i.e., more energetic) for distributions where the effective temperature is higher. If we restrict ourselves to the range marked V (*visible*) in Fig. 4.2, the stars must present colors from red to blue, corresponding to temperatures between approximately 3800 and 10 000 K.

The next issue is the *total energy* emitted by the star, since we now have an idea of how the photons are distributed. The black body emission problem remained unsolved until the first years of the 20th century, as discussed in Chap. 2. In fact, the functional shape of the curves in Fig. 4.2 corresponds to the expression (2.3), a consequence of the discrete nature of light. Note, however, that the *total flux* that emerges from a black body studied by G. Kirchhoff and others, i.e., the energy emitted per unit time and per unit area, has a very simple form: the result is proportional to the temperature to the fourth power, multiplied by a universal constant σ, and is completely independent of the composition of the body:

$$F = \sigma T^4 . \tag{4.1}$$

In particular, to calculate the total emerging flux it is enough to determine the stellar temperature by finding the maximum emission. However, the flux is not the most relevant quantity, since very different stars located at very different distances can lead to the same flux. Thus, an independent estimate of the *distance* is essential to convert the flux (relative) into luminosity (absolute), multiplying by the area of emission:

$$L = 4\pi R^2 \times \sigma T^4 . \tag{4.2}$$

Now we have assembled the basic framework that allowed E. Hertzsprung and H.N. Russell to propose the diagram that bears their names and serves to classify the stars. On the horizontal axis, we put the effective temperature of the emission, which is the one in (4.1) and (4.2), indicating in which region the above-mentioned maximum falls; and on the vertical axis, the luminosity (called *power* in Physics courses), a

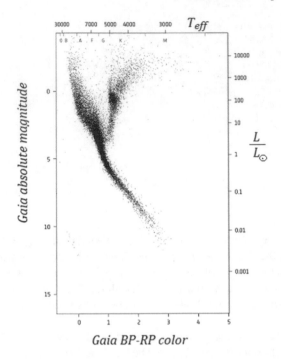

Fig. 4.3 Hertzsprung–Russell diagram built with data from the Gaia mission. Note that in addition to temperature and luminosity, we could have used other variables common in Astronomy, such as *absolute magnitude* and *spectral class* which are also equivalent to the former. For historical reasons, astronomers put the temperature scale growing in the direction of the origin, that is, the lowest temperatures are progressively farther from the origin. The location of regions containing stars and those lacking them must be explained by the theory of Stellar Evolution which we will study below. Credit: ESA/Gaia/DPAC; ESA/Hipparcos

measure of the total emitted energy which contains structural information of the star, viz., the radius R in (4.2), but which can be determined independently by measuring the flux and estimating the distance (the latter by various methods that feature different errors). Thus, the *HR diagram* in Fig. 4.3. can be constructed for all stars that emit as black bodies [2].

Inspection of the HR diagram shows a quite remarkable fact: there are populated and empty regions. Moreover, the vast majority of stars are located in a strip that goes from the top left to the bottom right. Of course, there has to be some important reason for the existence of this strip.

We can go even further without formulating any detailed description with the following observation. If we draw a horizontal line indicating *constant luminosity* in the upper half of the diagram, we see that there are stars for very different temperatures that have the same luminosity. From the generic expression (4.1), we see that this is only possible if the radii R are also very different, and in inverse relation to the temperatures, because this is the only way to keep the product $R^2 \times T^4$ constant. This justifies the denomination of *giants* and *supergiants*, depending on the value of the luminosity, found in the astronomical literature, and begs a theoretical explanation to understand why star radii are so different, even for the same star in different stages of its evolution. A vertical line T_{eff} = const. also reveals that there are stars with luminosity that differ by several orders of magnitude, but show the same temperature. This is only possible if the radius R is much larger in the more luminous stars. Thus, we see that there is a lot of information in the HR diagram and a lot of work to be done to achieve a physical description that explains how it is populated.

Before proceeding, we would like to point out another fact of importance: in Ancient Greece, Hipparchus saw and catalogued essentially the *same stars* observed in the sky today. The fact that stars have not undergone significant changes in over 2000 years shows that they are highly stable. More precisely, it shows that there is a very stable state of equilibrium that makes them last much longer than a few millennia. When we observe the night sky we are in the presence of a kind of "snapshot" of the star population, just as we could get of any human population in a very crowded place, for example in a public park. In the same image we would have examples of several human "evolutionary stages" (babies, young people, adults, and elderly people) and our task would be analogous to the study of human biology, which aims to understand the aging process. Fortunately, the laws of Physics apply to our problem, as we will discuss below.

4.3 Physical Description of Stellar Structure

The previous observation regarding the state of equilibrium and the fact that it can sustain stars for many millions of years leads to the question of the kind of equilibrium we are talking about. If we consider the case of a mass held up by a spring on the surface of the Earth (Fig. 4.4 right), the mechanical balance of the system is guaranteed by the condition $\mathbf{F}_{grav} = \mathbf{F}_{spring}$. But if we imagine now that the mass (in the form of a small cube) is an element of a fluid, the equivalent expression $\mathbf{F}_{grav} = \mathbf{F}_{press}$ points to two important considerations: first, the gravitational field is not "external" as in the case of the little cube on the Earth's surface, but rather it is the very distribution of fluid that produces the gravitation, whence a star is often called a self-gravitating fluid; and second, the whole fluid distribution is also responsible for the force that sustains the "little cube" fluid element, by producing a pressure that balances the gravitation.

When dealing with a self-gravitating fluid, the mechanical balance of forces is called a *hydrostatic balance*. Free of any other forces, it is well known that the fluid will adopt a spherical form (to minimize its free energy). Thus, the hydrostatic equilibrium equation can be obtained by considering concentric shells of thickness dr, where there is a pressure difference $P(r) - P(r + dr)$ between the base and the top of any given shell. On the other hand, the shell is subject to the gravitational force that pulls it towards the center of the star (Fig. 4.5).

Now we can use calculus to express the forces in a simple way: the force produced by the pressure difference is

Fig. 4.4 Analogy between a
mass sustained by a spring in
the Earth's gravitational field
and its continuous analogue,
where the cube is an element
of fluid that produces the
gravitational field, and the
pressure felt by it is also
produced by the distribution
of fluid as a whole

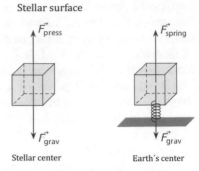

Fig. 4.5 Forces acting on a
spherical shell of
infinitesimal thickness dr

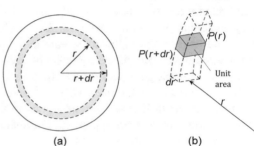

$$P(r) - P(r + dr) \approx -\frac{\partial P}{\partial r} dr \ , \tag{4.3}$$

and the gravitational pull is

$$\mathbf{F}_{\text{grav}} = -g(r)\rho(r)dr = -\rho\frac{GM(r)}{r^2}dr \ , \tag{4.4}$$

where it is clear that the local acceleration due to gravitation increases as one moves
outwards owing to the accumulation of shells within, and should be calculated using
the same density ρ of the fluid. The condition $\mathbf{F}_{\text{grav}} = \mathbf{F}_{\text{press}}$ leads immediately to

$$\frac{dP}{dr} = -\rho\frac{GM(r)\rho(r)}{r^2} \ . \tag{4.5}$$

Similarly, the continuity of mass indicates that, between radius r and radius $r + dr$,
there is a mass difference $dM = M(r + dr) - M(r) = 4\pi r^2 \rho dr$, i.e.,

$$\frac{dM}{dr} = 4\pi r^2 \rho(r) \ . \tag{4.6}$$

The set of Eqs. (4.5) and (4.6) would be sufficient for the description if stars did not
generate energy, but we know that this is not what happens in general. Stars radiating
like black bodies for many millions of years need a source of energy, and this energy

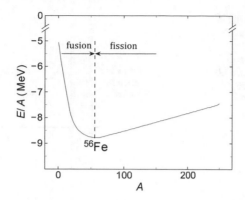

Fig. 4.6 Binding energy per nucleon as a function of the mass number A. The path followed by nuclear fusion is indicated *on the left*. Energy is gained when light nuclei merge, which is the mechanism used by stars. The fission of very heavy elements also delivers energy, since the fragments are more bound than the progenitor nucleus. This is the mechanism triggered in nuclear fission bombs (see [3] for more detail)

is related to the pressure that prevents collapse. Where exactly does this pressure come from?

A classical example with point particles shows us that it is possible to obtain energy by *binding* two or more particles that are initially in free states. In fact, if we suppose that two particles of masses m_1 and m_2 attract each other with Newton's gravitational force, they can form a bound state with total mass M given by

$$M = m_1 + m_2 - G\frac{m_1 m_2}{c^2 r} < m_1 + m_2 . \tag{4.7}$$

Therefore, the process of binding the particles results in a *lower* total mass than the initial one, since all the binding energies are negative. The excess mass multiplied by c^2 has to be expelled from the system into the environment in some form of energy, tending to increase the total energy and pressure of the environment.

The previous example serves to illustrate a generic fact that has its physical realization in the case of stars by means of *nuclear forces*. The accumulated empirical knowledge of the most common nuclei and the Periodic Table in general allow us to draw the important graph for stellar Astrophysics presented schematically in Fig. 4.6.

Since hydrogen is by far the most abundant element in the cosmos, and the easiest to fuse (the repulsion due to the electric charges is the smallest possible for any nuclei), let us start by studying what happens when two protons (hydrogen nuclei) collide. We begin by changing to the center-of-mass frame and transforming the two-body problem to a problem of a single body of reduced mass $\mu = m_1 m_2/(m_1 + m_2)$ in a central potential, a trick already used in standard courses on classical Mechanics. Note that in the case of two protons $\mu = m_p/2$, but the expression applies to any two particles. In our case, the potential is due to the electric charge everywhere except at

Fig. 4.7 Central potential in the problem of the collision of two protons, indicating the repulsive region due to the Coulomb interaction and the well where attractive nuclear forces dominate

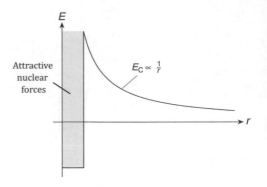

the closest range, where the (attractive) nuclear forces produce the "well" that allows the two protons to bind in a fusion process (Fig. 4.7).

Classically, only particles with energy higher than the top of the Coulomb barrier could fall into the attractive nuclear well region beyond. This far exceeds the average kinetic energy $\sim k_B T$ of the gas inside the Sun, and even the most energetic particles of the Maxwell–Boltzmann thermal distribution inside it. For this reason, Lord Kelvin sought the energy necessary to keep the Sun shining in the contraction of the solar structure. But with the formulation of Quantum Physics a few decades later, there was a way to justify the particles "crossing" the barrier with much lower energies, using the so-called *tunnel effect*. There was thus a non-zero probability of getting through, and the quantum formalism is used to calculate that. Here we will simply state that this probability is

$$P = \exp\left(-\frac{2\pi Z_1 Z_2 e^2}{\hbar v}\right),$$
(4.8)

where Z_1 and Z_2 are the particle charges, equal to unity for a proton (the hydrogen nucleus), and v is the relative velocity of the two particles measured in the center-of-mass system.

Defining the *cross-section* σ as the quotient of the number of reactions per particle, divided by the number of incident particles per unit area, all per unit time, and considering completely random collisions between the particles, the rate of fusion events r will be

$$r = N_1 N_2 v\sigma .$$
(4.9)

Note that this definition is not very rigorous, but suggests considering σ as the "effective area" in which fusion reactions can take place. In (4.9), when the "1" and "2" particles are the same, we obtain a factor N^2 typical of these processes, already discussed in elementary kinetic theory. The next step is to evaluate the number of particles that participate in the reaction between v and $v + dv$. Equation (4.9) actually describes one collision, but there are many others, and to take all of them into account we must integrate in order to sum over all velocities. As the quantity σ also depends on the velocity, the generalization we require is

$$r = N_1 N_2 \int_0^\infty v \times \sigma(v) \times \phi(v) \mathrm{d}^3 v \,, \tag{4.10}$$

where the classical Maxwell–Boltzmann distribution $\phi(v)$ gives the number of particles for each speed (or energy), and can be written as

$$\phi(v)\mathrm{d}^3 v = \left(\frac{\mu}{2\pi k_B T}\right)^{3/2} \mathrm{e}^{-\mu v^2/k_B T} 4\pi v^2 \mathrm{d}v \,, \tag{4.11}$$

where $4\pi v^2 \mathrm{d}v$ reflects the fact that we assume isotropy of the distribution in velocity space, i.e., independence of direction.

Now we need to take into account the probability that a particle of effective mass μ can cross the Coulomb barrier, as described by (4.8) which must multiply the integrand before the integration is done—this factor is not yet present in (4.10). But first we must change variables from speed to energy, as the independent variable. We begin with the cross-section, which must have the dimensions of area. In the microphysical domain we can see that the only possibility is to have something like $\pi \times \lambda^2$, where λ is the de Broglie wavelength of the proton, proportional to $1/E$ for non-relativistic particles. By separating the exponential and the factor $1/E$, we can introduce the so-called *astrophysical factor* $S(E)$ to "hide" more things, but in the hope that this will be almost constant with energy. Hence,

$$\sigma(E) = S(E) \times \frac{1}{E} \times \mathrm{e}^{-b/E^{1/2}} \,, \tag{4.12}$$

where

$$b = \frac{2^{3/2}}{\hbar} \pi^2 \mu^{1/2} Z_1 Z_2 e^2 \,.$$

Inserting the probability factor (4.8), the total rate is

$$r = \left(\frac{2}{k_B T}\right)^{3/2} \frac{N_1 N_2}{(\mu_m \pi)^{1/2}} \int_0^\infty S(E) \times \exp\left[-\left(\frac{E}{k_B T} + \frac{b}{E^{1/2}}\right)\right] \mathrm{d}E \,. \tag{4.13}$$

The rate of reactions results from an integrand that has two exponential functions with opposite behavior: if the energy increases, the exponential factor $\mathrm{e}^{-b/E^{1/2}}$ decreases, which is good because the tunnel effect makes it easier to "fall into the attractive well". But at the same time the number of particles with higher energies decreases according to the first exponential $\mathrm{e}^{-E/k_B T}$. If the energy is low, the opposite happens: there are many particles, but crossing the barrier is very difficult for them. Thus, there is a compromise between these two factors that indicates that both will be important only in the neighborhoods of a certain energy which optimizes the value of the integral. Below or abov this value the rate is virtually zero. This is the so-called *Gamow peak* (Fig. 4.8). The "optimal" energy E_0 can be evaluated and results in a value much greater than $k_B T$, i.e., much higher than the energy of most of the

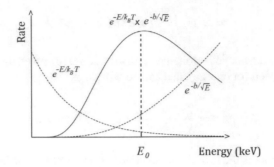

Fig. 4.8 Region around the Gamow peak. The factors in (4.13) behave in opposite ways and the result of the integral which determines the reaction rate is substantially different from zero only around the optimal energy E_0. Note that the former is generally beyond the peak of the Maxwell–Boltzmann maximum

Fig. 4.9 Fusion of hydrogen into helium as presented in many basic texts. The reality is much more complex and interesting than this figure would suggest (see text)

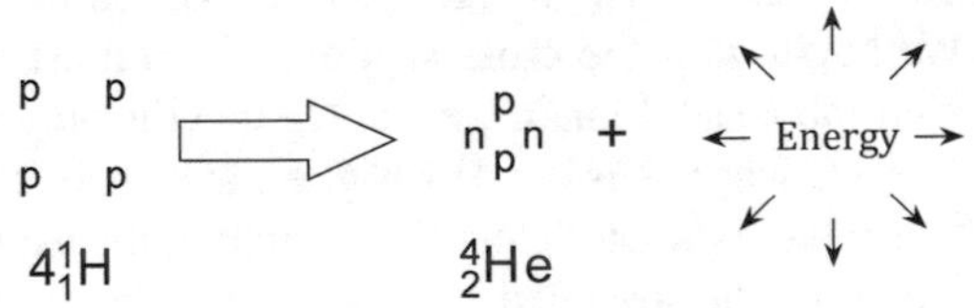

particles. The details of the integration are mathematically rather complicated, but the important thing to remember is that, after evaluating how much energy is released by the fusion of the two protons, the result can always be expressed in the form $\varepsilon(\rho, T) = \text{const.} \times \rho^{\alpha} T^{\beta}$, where ε is the energy released per gram of material per unit time. This is the form that appears in textbooks and which stems from a simple mathematical trick [4].

Before we continue with the formulation of the mathematical problem of stellar structure, it is necessary to point out some important characteristics of hydrogen fusion that are strongly determinant for Stellar Evolution. Normally, the exact form of the transformation of hydrogen into helium is not discussed in textbooks. Rather, a scheme of the type shown in Fig. 4.9 is presented, which is over-simplified and hides the true nature of the fusion path.

To begin, let us focus on the first two protons that merge. A plain, simple fusion is impossible. This is due to the fact that two protons *do not* have a bound state (i.e., there is no bound "diproton"). Only one proton and one neutron have a bound state (the deuteron, or deuterium nucleus written as ^2_1H, i.e., one proton and one neutron). Thus, the reaction involving the four protons could not even begin if one of the colliding protons did not become a neutron at the time of the collision. This process, governed by the weak interactions, can be thought of as the formation and immediate decay $^2_2\text{He} \rightarrow ^2_1\text{H} + e^+ + \nu_e$, where the positron e^+ and the neutrino ν_e escape the reaction zone. This must be very rare, and in any case most of the time the diproton ^2_2He, which is not bound, breaks down into two free protons again. Thus, only in a system with a gigantic number of collisions can a certain minimal number

$$\left.\begin{array}{c}
^{12}\text{C} + \text{p} \rightarrow {}^{13}\text{N} + \gamma \\
^{13}\text{N} \rightarrow {}^{13}\text{C} + e^+ + \nu_e \\
^{13}\text{C} + \text{p} \rightarrow {}^{14}\text{N} + \gamma \\
^{14}\text{N} + \text{p} \rightarrow {}^{15}\text{O} + \gamma \\
^{15}\text{O} \rightarrow {}^{15}\text{N} + e^+ + \nu_e \\
^{15}\text{N} + \text{p} \rightarrow {}^{12}\text{C} + {}^4\text{He}
\end{array}\right.$$

Fig. 4.10 Fundamental branch of the CNO cycle. A series of proton captures and decays on carbon, nitrogen, and oxygen culminates in the production of $^4_2$He from four "consumed" protons. The carbon is restored in the last stage, and the resulting helium formed from the four protons that enter the reactions on the left

of collisions produce the initial fusion. If we consider the probability that exactly at the time of the collision a proton decays into a neutron, this would happen once in 7×10^9 yr. In other words, every second, *one* out of 2×10^{17} collisions will produce a deuterium nucleus, while all others will result in two protons as before. This is a major bottleneck not captured at all by simple pictures like the one in Fig. 4.9.

If we focus on the lucky deuterium that survived, the later sequence is a little easier, the deuteron captures an additional proton and produces $^3_2$He, a reaction that is followed 86% of time by the capture of another proton that generates the final $^4_2$He. This sequence is known as the p-p I cycle. There are two other possibilities, today well understood, but the outcome is the same: conversion of four protons into a helium nucleus. As a corollary of this discussion, we come upon a very important fact: stars live for billions of years fusing hydrogen, a fact that would be impossible if two protons had a bound state. This last hypothetical state would cause the stars to explode in a few seconds, since only strong interactions would play a role in fusion, without any need for decay. However, as this is not the case, it is the weak interactions that determine the lifetimes of the stars, and it is not by chance that the above-mentioned rate of once in 7×10^9 yr is of the same order as the time the stars reside in the Main Sequence: almost no collisions will produce fusion, and the whole process is regulated by the slow release of energy, instead of the explosive situation that could have happened. We could say that we are "children" of inverse beta decay, because without it there would be no Stellar Evolution, no heavy elements, and no biology to use them and produce living organisms [5].

To complete this discussion, it is important to point out that there is another way to fuse hydrogen into helium which requires the presence of heavier elements such as carbon and oxygen and is called the *CNO cycle*. The CNO cycle is catalytic, because the heavy elements are not consumed, but enter reactions and return untouched to the environment. One of the branches (there are others of less importance) of the CNO cycle is shown in Fig. 4.10 [6].

The importance of the CNO cycle lies in its dependence on temperature: while the p-p cycle is weakly dependent on T, the CNO cycle grows very quickly with it. Calculations indicate an exponent $\beta = 1/4$ for the p-p cycle, but $\beta \approx 16$ for the CNO cycle, in the parametrization pointed out above for the rate of energy release

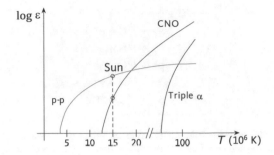

Fig. 4.11 Comparison of the energy generation between fusion in the p-p cycle (*green*) and the CNO cycle (*red*). In the Sun, the former is dominant, but for stars that have twice the solar mass or more, the CNO cycle dominates, since the central temperatures exceed about 1.8×10^7 K. The possibility of carbon fusion via the triple-α process occurs only for much higher temperatures

$\varepsilon(\rho, T)$. This means that, as we consider stars of larger masses, with increasing central temperatures, the CNO cycle will inevitably surpass the p-p cycle in importance. Since dependence on the former is much more pronounced, stars that have the CNO as their main mechanism of hydrogen fusion will expend energy much faster and spend less time in the Main Sequence. The situation is shown in Fig. 4.11 and the mass for which the conditions are reached is found to be slightly higher than $2M_\odot$. Thus, we have a physical reason to consider two branches in the main sequence: the upper MS ($M > 2M_\odot$) and the lower MS ($M < 2M_\odot$), to which the Sun belongs. The differences between the evolution of stars on the two branches will be discussed later.

We can now resume our description and write, in an analogous way to equations (4.5) and (4.6), an equation that relates the luminosity produced in a concentric shell layer of thickness dr to the rate of energy release ε. Within this shell the luminosity dL increases with the energy release according to

$$dL = \varepsilon dr \ , \tag{4.14}$$

and as before, the mass and radius differentials are related by $dM = 4\pi r^2 \rho dr$, yielding

$$\frac{dL}{dr} = 4\pi r^2 \rho \varepsilon \ , \tag{4.15}$$

which is coupled to (4.5) and (4.6) by the presence of the variable ρ and requires the calculation of the nuclear energy release per unit mass ε on the right-hand side. But there is another consideration regarding the release of energy in the stellar interior: it is clear that this energy will *move away* from the initial point, thus establishing a temperature profile: the only variable that does not yet appear explicitly in the set of equations is precisely the temperature. Thus, it is reasonable to ask ourselves whether this spatial distribution of temperature can be constructed from a fourth equation that features the temperature gradient d$T/$dr.

An initial hypothesis for energy mobility would be that the flux of radiation leaving the star from the inside, this being the main mechanism of energy transport, would depend on this temperature gradient, as happens in any diffusive process. We assume that the radiation collides very frequently with matter, so that each photon follows a long and tortuous path out to the surface. Note that the stars are *not* in thermodynamic equilibrium, since if they were they would not radiate net energy into space. But on small scales, there is a local balance between radiation and matter which establishes the temperature distribution $T(r)$ we are seeking, governed by gravitation and the other ingredients inside.

Consider once again a shell with thickness dr. The difference in the radiation flux that comes in through the base and the one that comes out through the top is

$$\sigma(T + dT)^4 - \sigma T^4 \approx 4\sigma T^3 dT , \qquad (4.16)$$

where we have *linearized* the Stefan–Boltzmann law, assuming that the temperature difference is small. This difference across the layer can be written as $dT = \bar{\lambda} dT/dr$, hence, proportional to the temperature gradient, where the mean free path of the photons denoted by $\bar{\lambda}$ is a quantity that contains an energy average over all the photon–matter interaction processes, since it corresponds to a diffusion process. Inserting this, the flux F is obtained in the form

$$F = -4\bar{\lambda}\sigma T^3 \frac{dT}{dr} . \qquad (4.17)$$

A more detailed statistical treatment shows that the mean free path can be replaced by another variable $\bar{\kappa}$ called the *opacity*, where $\bar{\lambda} = (\bar{\kappa}\rho)^{-1}$. Furthermore, the flux F is related to the luminosity in the layer by $F = L/4\pi r^2$. Note that this is not to be confused with the total luminosity of the HR diagram; it is the energy per unit of time that passes through the inside layer. This latter quantity is the same as the flux expressed in (4.17). Substituting in and extracting the temperature gradient dT/dr, we obtain [6]

$$\frac{dT}{dr} = -\frac{3}{16} \frac{\bar{\kappa}\rho}{\sigma T^3} \frac{L}{4\pi r^2} . \qquad (4.18)$$

In much the same way that we calculate the rate of nuclear reactions $\varepsilon(\rho, T)$ and then plug it into the structure equations, we can do the same for the opacity $\bar{\kappa}$, which involves several physical processes that are more or less relevant for each range of densities and temperatures. Four basic processes contribute to the opacity: bound–bound processes (photoabsorption), bound–free processes (photoionization), free–free processes (reverse bremsstrahlung), and Compton scattering, all of which are shown graphically in Fig. 4.12.

The full calculation of opacities is complicated and requires the tools of Quantum Mechanics. However, the averages for the first three can be expressed as $\bar{\kappa} \propto \rho T^{-3.5}$, the so-called *Kramers form* (although the coefficients turn out to be very different). Compton scattering is a process that, at least in the limit of low energies (generally

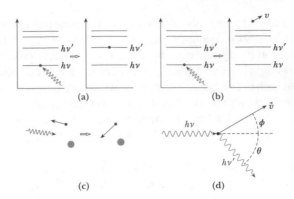

Fig. 4.12 The four basic processes that contribute to stellar opacity. Note that all of them are ultimately "obstacles" to the outward progress of the photons, which are forced to exchange energy and momentum with an electron, whether it is bound, as in cases (**a**) and (**b**), or free. In the first three, the initial photon disappears, since they are *absorption* processes. In Compton scattering, already discussed in Chap. 2, the photon is *scattered* and changes direction with a change in its initial energy

applicable to stars), is independent of frequency, and thus results in a constant value $\bar{\kappa}_{\text{Compton}} = 0.2(1 + X)\,\text{cm}^{-2}\text{g}^{-1}$. The most complete treatment of this subject can be found in [6].

Equation (4.18) completes the system of equations as long as the energy transport proceeds in a diffusive mode throughout the star, with photons colliding with matter driven by the (small) temperature gradient that allowed us to linearize the system. But this diffusion can be overcome if the gradient exceeds a critical value. This transition is well known in everyday life: if we put water to boil, well before this happens, the water in the pot will begin to show large scale fluid movements (Fig. 4.13), since it is much more efficient to transport heat this way. This analogy is quite close to what happens in a star, and illustrates the establishment of (stationary) *convection*. If we continue to supply enough heat, this regular convection will give way to turbulent movements, involving eddies of various length scales, which are even more efficient for transporting energy (and more difficult to describe mathematically).

The value of the so-called *adiabatic gradient* separating the diffusive and convective regimes is calculable, with a value of 2/5 for a monatomic gas. In other words, it is not possible for the gradient dT/dr to grow indefinitely: when the adiabatic value is reached, convection takes over (this is called the *Schwarzschild criterion*). From this point on, we cannot use (4.18) as-is: the second expression must be replaced by one that is valid for convection. One crude approximation assumes that the gradient will stay at its adiabatic value. After all, that is why convection is efficient and flattens the temperature gradient, making the temperature practically constant. Then, (4.18) will be replaced by

$$\frac{dT}{dr} = \left(1 - \frac{1}{\gamma}\right)\frac{T}{P}\frac{dP}{dr}\,, \tag{4.19}$$

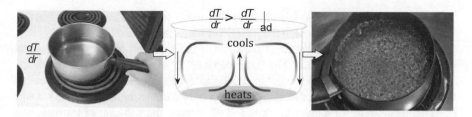

Fig. 4.13 Putting a pan on the stove is similar to heating the base of a star layer. The temperature gradient dT/dr will grow until it exceeds the so-called adiabatic gradient $dT/dr|_{ad}$, and the convective instability starts to produce regular movements (*central panel*). If we continue to increase the gradient, there will eventually be a transition to the turbulent regime (*right panel*)

where on the right-hand side we have used the very definition of the adiabatic gradient $dT/dr|_{ad}$. This is perhaps too crude, and another improvement widely used in Astrophysics is the so-called *mixing-length theory*, where the hypothesis that convection carries heat by means of a "typical" convective bubble leads to a non-linear expression of the flux. Finally, there are other models for calculating the convective flux using a full distribution of bubble sizes, and these are even more complicated. There will be no need to go into all the details of this modeling here, but it is important to remember that Schwarzschild's criterion $dT/dr < dT/dr|_{ad}$ must be monitored layer by layer in the construction of the star model, and when it is not satisfied, (4.18) must be replaced by (4.19) or something better in order to describe the convective region.

4.4 Some General Considerations about Stellar Evolution

The system of coupled equations (4.5), (4.6), (4.15), and (4.18) [or (4.19) if the convective regime has been reached], with the stellar boundary conditions

$$M(r = 0) = 0 , \tag{4.20}$$

$$L(r = 0) = 0 , \tag{4.21}$$

$$P(r = R) = 0 , \tag{4.22}$$

$$T(r = R) = 0 , \tag{4.23}$$

are the set that allows us to generate the stellar models that we will present below, then compare their characteristics with real star observations. In other words, a star model is the solution yielding the four functions $P(r)$, $M(r)$, $T(r)$, and $L(r)$, satisfying a constraint $P(\rho)$ called the *equation of state* and giving the constitutive functions $\bar{\kappa}$ and ε. The latter are functions of temperature and density. The last condition may seem a little strange since we should impose $T(r = R) = T_{eff}$, but this greatly complicates the numerical calculations and the form (4.23) is preferred. The effective temperature can then be calculated *a posteriori* when the model is solved.

As with any other mathematical problem, it is important to know the general characteristics of the solutions. The "well behaved" solutions are unique and mathematically stable. In fact, in the theory of Stellar Evolution, there is a result called the *Russell–Vogt theorem*, although it is in fact only a conjecture, since it has never been demonstrated: for a given mass and composition, the solution of the structure equations is unique. With some reservations, we can therefore go ahead with the guarantee of being in the presence of a well formulated problem with physically acceptable solutions.

We are now in a position to discuss the problem of stellar stability raised at the beginning of this chapter. We begin (4.5) and (4.6), which ensure the hydrostatic (mechanical) equilibrium. A simple manipulation of (4.5) consists in multiplying both sides by the volume $V = 4\pi r^3/3$ and integrating over the mass, i.e.,

$$\int \frac{4}{3}\pi r^3 \frac{\mathrm{d}P}{\mathrm{d}M}\mathrm{d}M = -\frac{1}{3}\int \frac{GM}{r}\mathrm{d}M , \qquad (4.24)$$

bearing in mind that $\mathrm{d}/\mathrm{d}M = (\mathrm{d}/\mathrm{d}r) \times (\mathrm{d}r/\mathrm{d}M)$. The expression on the right is recognized as the gravitational energy of the star $\Omega_G = -\int (GM/r)\mathrm{d}M$ and the left-hand side can be integrated by parts to yield

$$\int_{P_C}^{P_S} V\mathrm{d}P = PV\Big|_{P_C}^{P_S} - \int_0^{V_S} P\mathrm{d}V . \qquad (4.25)$$

Here the first term on the right is zero because, on the surface S, the pressure is zero, and in the center the volume V is zero. Thus, we have $\Omega_G = -3\int_0^{V_S} P\mathrm{d}V$. If we transform the integral from the volume to mass using the fact that $\mathrm{d}V = \mathrm{d}M/\rho$, the general result is

$$\Omega_G = -3\int_0^M \frac{P}{\rho}\mathrm{d}M . \qquad (4.26)$$

However, in the specific case of an ideal gas, $P = nk_BT = (\rho/\mu m_H)k_BT$, and we can identify the *internal energy u* per unit mass as the average kinetic energy per particle $3/2k_BT$ divided by the mass (μm_H, where μ is the average molecular weight). Thus, $u = 3P/2\rho$, and the Virial theorem for an ideal gas is

$$\Omega_G = -2U , \qquad (4.27)$$

with $U = \int u\,\mathrm{d}M$ the total internal energy of the star.

The Virial theorem contains the physical basis required to explain in simple terms how a star "works". The first important thing is that, in hydrostatic balance where there is no large-scale movement of the gas, the total energy of the star is $E_{tot} = \Omega_G + U$. The Virial relationship tells us immediately that $E_{tot} = \Omega_G/2 < 0$. The star is *bound* (negative total energy) and distributes its energy according to (4.27) as a result of the microscopic exchange of this energy between particles. Moreover, if we differentiate, we obtain $\dot{\Omega}_g = -2\dot{E}_{tot} < 0$, where $\dot{E}_{tot} = L$ is the luminosity, i.e., the loss of radiative energy per unit time. This means that when the star loses energy, it contracts, and since $L > 0$, we have $\dot{E}_{tot} > 0$ and the contraction makes the gas heat up. Of course, one would not normally expect an object that loses energy to heat up, and this emphasizes the important role of gravitation in stellar equilibrium, justifying the (somewhat cumbersome) statement in the literature that stars have a *negative specific heat*.

It should be pointed out that, although we have began our derivation of the Virial theorem with the hydrostatic equilibrium equation (4.5), in fact this Virial energy sharing precedes the establishment of such an equilibrium. In other words, a system in hydrostatic equilibrium always satisfies the Virial relationship (4.27), while the converse is not true. A system can be "virialized", that is, sharing the energy according to (4.27), without being in hydrostatic equilibrium—proto-stars in formation are a good example of this. Of course, the total energy must also be conserved. We will see later on the importance for the evolution of the stars of maintaining both the Virial relationship and energy conservation at the same time.

To conclude, and thinking ahead to the key features of Stellar Evolution discussed below, we define some timescales that will be important for the question of stellar stability and which are implicit in the previous structure equations. Comparisons between them will allows us to understand how stellar structure reacts to the perturbations that take it out of equilibrium.

The first important timescale is the *dynamical timescale*. Physically, it is the time for a sound wave to cross the stellar radius. It is practically the same time that it takes a particle on the surface to "fall" to the center in an idealized way. Numerically, we can use conservation of energy, imagining the whole mass to be concentrated at the center:

$$\tau_{dyn} \approx \sqrt{\frac{R^3}{GM}} = 0.02 \left(\frac{R}{R_\odot}\right)^{3/2} \left(\frac{M}{M_\odot}\right)^{1/2} \text{ days} . \tag{4.28}$$

That is, in a few hours or so, a star can restore the hydrostatic balance, since the fluid is in fast communication. It is hard to imagine the hydrostatic balance being violated in stars, regardless of the type of perturbation.

The second timescale is the so-called *thermal* or *diffusive* timescale, telling us how long it takes for the star to establish a stationary temperature distribution if perturbed. Assuming that diffusion of photons is the dominant mechanism, with coefficient of diffusion D, we have

$$\tau_{diff} \approx \frac{R^2}{D} = 10^5 \frac{\kappa}{\kappa_\odot} \frac{\bar{\rho}}{\bar{\rho}_\odot} \left(\frac{R}{R_\odot}\right)^2 \text{ yr} , \tag{4.29}$$

where we have used the fact that $D = c/\kappa\bar{\rho}$ for the diffusion of photons through the stellar material. This is much longer than τ_{dyn}, although quite short by astronomical standards.

The third timescale is the *Kelvin–Helmholtz time*, corresponding to the time it takes the star to radiate a substantial fraction of the energy contained in it. From the above discussion of the Virial theorem, we have $L \sim -\dot{\Omega}_g/2$, i.e.,

$$\tau_{K-H} \approx \frac{\Omega_G}{2L} = \frac{GM^2}{2RL} = 1.5 \times 10^7 \frac{R_\odot}{R} \frac{L_\odot}{L} \left(\frac{M}{M_\odot}\right)^2 \text{ yr} . \tag{4.30}$$

The fourth and last timescale is the *nuclear timescale*, the time during which the fusion reactions maintain the brightness of the star. If in each reaction a fraction $\xi \sim 0.007$ of the rest mass is released (this number is valid for H $\rightarrow$ He), and a fraction $f \geq 0.5$ of the star's mass is available for fusion, the nuclear timescale is

$$\tau_{nuc} \approx f\xi \frac{Mc^2}{L} = 10^{10} \frac{M}{M_\odot} \frac{L_\odot}{L} \text{ yr} . \tag{4.31}$$

Thus, the required fundamental hierarchy, the one that makes stars stable and very long-lived, is

$$\tau_{dyn} \ll \tau_{diff} \ll \tau_{K-H} \ll \tau_{nuc} . \tag{4.32}$$

We see that the time that governs the lifetime of a star is the nuclear time (more precisely, the weak interaction decay time described above as a "bottleneck" for fusion reactions), and stars can be considered in hydrostatic and thermal balance in any situation. This is why the lifetime of the stars is so long. Only when some of these inequalities are violated during a star's evolution will there be important changes. Therefore, we may state that the life of a star consists of *long periods* of stationary equilibrium separated by moments when the hierarchy in (4.32) is violated and the star is forced to seek a new stationary state or phase in its evolution [4].

4.5   Stellar Evolution: Low Mass Stars

What we consider low or high mass stems from the different possible types of physical behavior that we will now define. We have already pointed out that around $2M_\odot$, the CNO cycle begins to dominate the p-p, whereupon the production of nuclear energy is greatly accelerated due to its stronger dependence on temperature. This defines the point of separation between the stars of the *lower Main Sequence*, to which the Sun belongs, and those of the *upper main sequence*. But there is another important limit around $8M_\odot$: after the ignition of helium to produce carbon, stars of lower mass will never be able to fuse the resulting carbon, so their evolution stops there. Those with higher masses (especially above $\sim 10M_\odot$) go through a succession of nuclear cycles right up to the burning of silicon and are called high mass stars. The less massive ones, with $M \leq 8M_\odot$, are thus said to be low mass, with evolution and fate similar to our Sun [2, 4, 6].

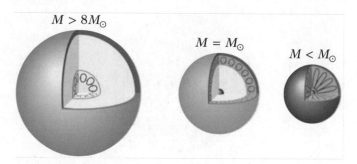

Fig. 4.14 Visual comparison between three stars, showing the convective and radiative zones. The smallest, up to ~ $0.3M_\odot$, shown *on the right*, is fully convective from the outside in. This convective envelope recedes for higher masses (*center*). The high mass star produces almost all its helium through the CNO cycle, the temperature profile is much more abrupt and as a consequence it is the central core that becomes convective. Around $20M_\odot$ the convective zone occupies almost the entire star from the inside out

The life of a star begins effectively when a cloud of gas contracts and the center of the cloud reaches density and temperature conditions sufficient to ignite the hydrogen fusion reactions discussed previously. This stage is never actually reached for masses smaller than $0.08M_\odot$ (about 85 times the mass of Jupiter), so these brown dwarfs will never convert to stars and will only emit a small part of their original energy content while cooling indefinitely. The core of stars above this lower limit will be totally convective up to about $0.3M_\odot$, at which mass they begin to satisfy Schwarzschild's criterion in the center and the fraction of the mass occupied by this radiative core grows for stars of progressively higher masses. For the Sun, this radiative core extends to about 70% of the radius, and only the envelope undergoes convection, as shown in Fig. 4.14.

Low-mass stars spend several billion years located almost at the same point in the HR diagram. Their lifetime can be estimated empirically, knowing that there is a mass-to-light relation for the Main Sequence, viz., $L \propto M^{3.5}$. As the basic estimate is simply $\tau_{MS} \approx E_{nuc}/L$, and expressing all in terms of solar quantities, we have immediately

$$\frac{\tau_{MS}}{\tau_\odot} = \left(\frac{M_\odot}{M}\right)^{2/5}. \tag{4.33}$$

In other words, very low mass stars will continue on the Main Sequence for many billions years, while more massive ones should live a few tens of millions in this situation. (Note that this relationship is also valid for high mass stars, those with shorter lives on the MS.) The situation can be compared to what happens with a car and its fuel: an economical car depletes its fuel very slowly to run more kilometers (that is, it "lives" longer). But a sports car uses its fuel very quickly, because it is not made to do otherwise, so its "lifetime" is much shorter (until it is refuelled, something which the stars cannot do). This is what happens with stars as a function of their mass: they enjoy a long or short life in accordance with (4.33).

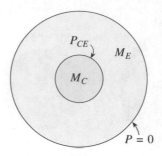

Fig. 4.15 Sketch of the core–envelope configuration for a low-mass star. The pressure P_{CE} should be the same on both sides of the core–envelope interface, but calculations indicate that the core reaches a maximum pressure beyond which it cannot remain in equilibrium with the envelope when only a small fraction of the total original hydrogen has been consumed by the fusion

Life on the Main Sequence does not provoke any notable structural changes in the star, but rather a slow differentiation of the central core which is enriching itself with helium while being depleted of hydrogen. So how long can a star go on fusing hydrogen and maintain itself in the MS? The answer seems evident, but it hides surprises. In high-mass stars, the convective core homogenizes the composition and fusion ends when all the hydrogen is used up, as would be expected. But in low-mass stars, the central core is changing composition because helium is produced, precisely from the fusion that consumes the hydrogen. Thus, the core is gradually becoming different from the envelope (Fig. 4.15). The presence of increasing amounts of inert helium in the core produces an unsustainable situation for this configuration in the long run: the pressure inside the core must support the envelope situated above it, i.e., the pressure P_{CE} must be the same on both sides of the core–envelope interface. But as this pressure consists of two terms with different signals, it has a *maximum* as a function of the total mass of the core M_C. When this maximum is reached, by constant addition of helium, hydrostatic equilibrium is no longer possible and the star must seek a new stationary equilibrium because it is no longer possible to support the envelope on top.

As a function of the composition, this condition can be written as

$$\frac{M_C}{M_*} = 0.37 \left(\frac{\mu_E}{\mu_C} \right)^2 , \qquad (4.34)$$

where μ_C and μ_E are the molecular weights in the core and envelope, respectively. Numerically, the core can no longer support the envelope when $M_C \sim 0.1 M_*$, or when about 13% of the star's original hydrogen has been consumed. Therefore, there is still a lot of hydrogen in the envelope of a low-mass star when the fusion stops, in contrast to the high-mass case. This result is called the *Schoenberg–Chandrasekhar limit*, not to be confused with the *Chandrasekhar limit*, which refers to something else (see Chap. 6). It shows that the reason for low-mass stars exiting from the MS is structural, due to the impossibility of remaining in hydrostatic equilibrium, and is *not* due to the exhaustion of hydrogen in the star. In fact more than 80% of the original hydrogen will still be present in it, although all outside the central core [4]. This contribution by the Brazilian scientist Mário Schoenberg is of great importance in the context of Stellar Evolution, and is among the best results produced by the then young University of São Paulo in 1942.

Taking into account the fact that the (thermal) energy content is still very large inside the star, there must be a quasi-hydrostatic readjustment, that is, without collapse, controlled by the Kelvin–Helmholtz timescale. In other words, the star is in search of a new stable equilibrium, as we said before, since the inequalities (4.32) are violated at the moment of reaching the Schoenberg–Chandrasekhar limit. The core is now inert (with no nuclear reactions) and the p-p (or CNO) cycle does not really stop, but occurs in a spherical shell around this inert core. At this point, the envelope *expands greatly*, while the core contracts and heats up. What is the physical reason for this behavior, and what explains the star's displacement in the HR diagram?

A quantitative and consistent explanation can be found by considering that *both* the Virial ratio (4.27) and the total energy must be conserved simultaneously. For times $\tau \gg \tau_{\text{dyn}}$, the Virial energy distribution leads to $\langle \Omega_G \rangle + 2\langle U \rangle = 0$, where $\langle \ \rangle$ indicates the spatial average in the star composed of the core plus envelope. On the other hand, conservation of the total energy requires

$$\langle \Omega_G \rangle + \langle U \rangle - \int_0^t L \, \mathrm{d}t + \int \int_0^t \varepsilon \, \mathrm{d}t \, \mathrm{d}V = \text{constant} . \qquad (4.35)$$

Moreover, for $\tau \gg \tau_{\text{dyn}}$, the two integrals tend to zero, so $\langle \Omega_G \rangle + \langle U \rangle = 0$, recalling that, according to (4.28), τ_{dyn} is very short. In addition to the Virial expression $\langle \Omega_G \rangle + 2\langle U \rangle = 0$, we know that $\langle \Omega_G \rangle$ and $\langle U \rangle$ should be constant separately. Now we can write explicitly that

$$\Omega_G = \frac{GM_C^2}{R_C} + \frac{GM_C M_{\text{env}}}{R_*} = \text{constant}' ,$$

to obtain the variation of the total stellar radius R_* with the radius of the core R_C, by differentiating this last expression and rearranging the factors:

$$\frac{\mathrm{d}R_*}{\mathrm{d}R_C} = -\frac{M_C}{M_{\text{env}}} \left(\frac{R_*}{R_C} \right)^2 . \qquad (4.36)$$

The stellar radius increases as the core radius shrinks (negative sign) amplified by the large squared factor. As a result of this expansion the star moves, with little change in luminosity, to lower effective temperatures, since if $L = \sigma T^4 \times 4\pi R^2$ is practically constant and the radius R needs to grow as indicated in Fig. 4.16, the effective temperature T needs to decrease to ensure the constancy of the product. This happens until the surface is cold enough to allow the formation of the hydrogen ion H^-, technically an anion with two electrons and a proton. The key issue is that the opacity of the H^- ion is very high, and this therefore amounts to "stopping"

Fig. 4.16 Comparison of the Sun, Pollux ($1.86M_\odot$, red giant), and Arcturus (around $1.5M_\odot$, red giant), quantifying the expansion of the stellar envelope visually in this phase, as indicated by (4.36). Credit: Arcturus by Pablo Carlos Budassi

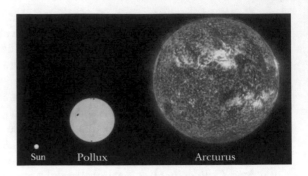

Fig. 4.17 Path of a star like the Sun in the HR diagram. After exit at point (a), the expansion of the envelope permits the formation of the H⁻ ion at point (b) and the trajectory becomes almost vertical towards point (c) for the physical reasons explained in the text

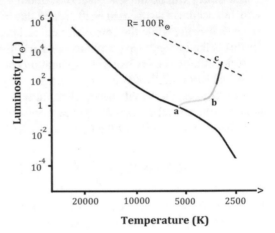

the flow of radiation from the inside at the temperature at which it forms. As a consequence, the star *increases* in luminosity L, rising through the *giant branch* (Fig. 4.17, trajectory in red).

Eventually, this path takes the star to the top of the giant branch, marked "c" in Fig. 4.17, where its radius is 100–200 times the radius it had in the Main Sequence, and its luminosity is about 1000 times greater. At this point the material in the core of a low-mass star, which has continued contracting and heating during this process, generally becomes *degenerate*, i.e., it is no longer an ordinary gas: the density is so high that the dominant component of the pressure is due to the electrons satisfying Pauli's exclusion principle. Although we shall discuss degenerate gases in more detail in a later Chapter, we may establish its basic description now, since it is a very important concept for the evolution of stars.

We are accustomed to think that a gas has a pressure which increases with temperature. This is true for a classical gas, where thermal agitation is responsible for the pressure. However, if we consider matter of increasing densities, there will be a moment where electrons can only occupy energy states of a maximum of two per each "little cube" in the phase space spanned by the position variable x and the momentum variable p. This cube now has a volume of the order of the cube of the Planck constant, i.e., $(\Delta \times \Delta p)^3 \sim \hbar^3$. The gas changes regime and is no longer gov-

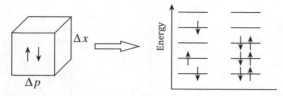

Fig. 4.18 *Left*: A little cube of the phase space where only two electrons per state can fit, with opposing spins. *Right*: A classical gas with a lot of accessible states gives way to a degenerate gas, where only two electrons can occupy each energy state

erned by Classical Physics, but enters the (quantum) *degenerate regime*. Basically, this means that, as we "squeeze" the electrons further, the principles determining the new source of pressure are the uncertainty principle, which determines the dimensions of the phase space, and Pauli's principle, which states that there can be no more than two electrons occupying one of the little cubes, and then only when they have opposite spin numbers. Meanwhile, thermal agitation is no longer important, but not because the temperature is low (it is very high when measured in K), but because degeneracy provides a much higher pressure. The situation is illustrated in Fig. 4.18.

We will soon see that the pressure can be calculated in a simple way in the new degenerate regime, since the pressure is just the derivative of the energy per particle with respect to the volume, that is, it describes how matter reacts if we try to compress it. The results, for the case of non-relativistic and ultra-relativistic electrons are:

$$P \rightarrow \frac{\hbar^2}{m} n^{5/3} \quad \text{(non-relativistic electrons)}, \tag{4.37}$$

$$P \rightarrow \frac{\hbar^2}{m} n^{4/3} \quad \text{(ultra-relativistic electrons)}, \tag{4.38}$$

where $n = N/V$ is the number of particles (electrons) per unit volume. From these expressions it can be checked that this degenerate pressure is of purely quantum origin: if the Planck constant were zero it would not exist. The passage from one regime to the other is shown schematically in Fig. 4.19.

Going back to our previous discussion, the most important fact is that the pressure of a degenerate gas *does not* depend on the temperature. When helium ignition begins in the core (above about 10^8K), the core will not "react", in the sense that the pressure will not increase by the release of helium fusion energy. This fusion, therefore, gets out of control, since without an increase in pressure the star will not expand to bring the temperature down. This phenomenon is called the *helium flash*, and releases an internal luminosity many orders of magnitude greater than the initial one. However, since this release is buried deep down and occurs very quickly, it does not produce pronounced observable consequences. In fact there is a "hunt" for the stars that may be undergoing the flash: those that are located at the top of the giant branch.

As a result of the thermal instability that leads to the flash, the star finally releases so much energy internally that it comes out of the degenerate regime. The pressure

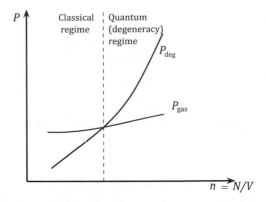

is once again dominated by the temperature and the star finally expands. This makes
it possible to control the helium nuclear reactions known as the *triple-alpha process*,
already shown on the right in Fig. 4.11. This helium fusion cycle also has "hidden
secrets". To begin with, it is a cycle that could not happen if it were necessary for three
helium nuclei to meet simultaneously, since this would be almost impossible. As with
the reaction of two protons with decay in the p–p cycle, the initial reaction is $2\,^4\text{He} \rightarrow$
^8Be, and at any instant, there is 1 beryllium nucleus for every billion helium nuclei.
However, the nucleus ^8Be is highly unstable, and in fact there are no stable elements
in the Periodic Table between $A = 5$ and $A = 8$. Thus, the beryllium nucleus decays
rather quickly, although it lives much longer than a simple random collision, and
there would therefore be time to capture a third helium, making $^8\text{Be}+^4\text{He}$, were it
not for the fact that this reaction is *prohibited* by the conservation of fundamental
quantities, in this case the parity of the states. It was F. Hoyle who reasoned that there
must be an excited carbon nucleus to complete the reaction, serving as a "doorway"
for the arrival and then decay into ordinary (ground state) carbon. Indeed, without a
doorway state the reaction would not occur and there would be no carbon from stars
(nor human beings made from it). This excited carbon state was found soon after.
When the decay is completed, we can write

$$^4\text{He} \rightarrow \,^{12}\text{C} + \gamma \,, \tag{4.39}$$

although it should be remembered that this is just a shorthand for something much
more complicated [6]. Performing the calculations to find the energy released by
each reaction, we arrive at the expression $\varepsilon(\rho, T) = \text{constant} \times \rho^\alpha T^\beta$, with $\alpha = 2$
and $\beta = 41$. The dependence on temperature is extreme! We may thus expect helium
exhaustion to occur much faster than in the case of hydrogen. Stationary fusion by
means of the triple-α process causes the star to "descend" from the giant branch
and establish itself in the so-called *horizontal branch* if its metallicity is low (or
red clump, if its metallicity is high). Here, the core produces carbon from helium,
while the fusion of hydrogen into helium continues in the surrounding spherical shell
(Fig. 4.20).

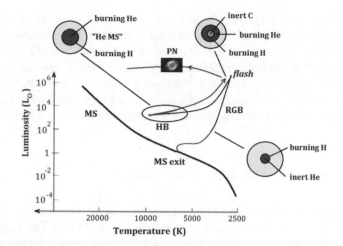

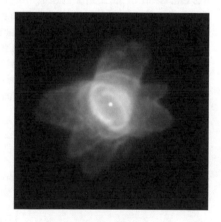

Fig. 4.20 Trajectories of a $1M_\odot$ star in the ascending and descending giant branch, to establish itself in the "helium Main Sequence", as indicated. Exit from the horizontal branch or red clump occurs for the same physical reasons as those discussed earlier for the Main Sequence, now applied to the helium fusion reaction

Fig. 4.21 The He 2-47 planetary nebula observed by the Hubble Space Telescope. Note the former C–O core shining as a point at the center. This must become a true white dwarf when it cools down. Credit: Hubble Space Telescope/NASA

In much the same way as the star's core reached the Schoenberg–Chandrasekhar limit to leave the MS, there is an analogous limit for the carbon-rich inert core. For the same physical reasons as already described, the star should now ascend to the so-called *asymptotic giant branch* (AGB). This time thermal instability occurs in the concentric shell, although it is not degenerate, and leads to *thermal pulses* which end up expelling the envelope and producing the beautiful images that we have of planetary nebulae (Fig. 4.21), leaving behind the carbon enriched core and also oxygen: the capture of α particles by carbon is inevitable, and becomes more important for higher masses, near the upper end of the range we are considering here, viz., $8M_\odot$.

   This leftover C–O core, initially very hot, will cool down over several Gyr to become a cold white dwarf, something we shall study in a following chapter. Finally, we would like to point out that there is observational evidence of white dwarf production by stars that had higher mass than $7.5 M_\odot$ in the Main Sequence. In the range closest to this upper limit, the cores suffer more α captures and end up with a composition of C–O in approximately equal parts (and in a degenerate state, even before the thermal pulses). Heavier compositions are possible for the most extreme WDs.

4.6   Stellar Evolution: High Mass Stars

We have discussed the evolution of low-mass stars, and we can now turn to the high-mass ones, already defined as those that exceed roughly $8 M_\odot$. The dividing line between the two groups is given by the mass where carbon cannot be fused because the temperature in the central region does not reach the ignition value of around 8×10^8 K. But it is important to emphasize that what we have here is a completely new ingredient in stellar evolution, and one that effectively prevents this ignition: the emission of neutrinos that is triggered at temperatures of this order, and cools the region that could otherwise fuse very efficiently. Thus, in order to fuse carbon at these temperatures, the reaction rate must exceed the core's neutrino emission rate, a condition that recurs for all the fusion reactions that follow. Moreover, thanks to the neutrinos, the cores and envelopes now follow different paths, as though they can no longer see each other. Technically, we speak of *thermal decoupling*. The thermally decoupled core, supported by electrons is degenerate and to a good approximation satisfies the condition [6]

$$\frac{T^3}{\rho} = \text{constant} , \tag{4.40}$$

for core densities $\rho \approx 10^5$ g cm^{-3}. Then, provided that the mass of the star is sufficient, the "inert" carbon that has accumulated finally fuses according to the reaction

$$^{12}\text{C} + {}^{12}\text{C} \rightarrow \left({}^{24}\text{Mg}\right)^* , \tag{4.41}$$

where the excited magnesium nucleus $\left({}^{24}\text{Mg}\right)^*$ decays in many different ways (remembering the case of the excited carbon proposed by Hoyle) which need to be added together to obtain the final reaction rate proportional to T^{29}. This cycle of carbon fusion lasts substantially less than the triple-α, and when exhausted it makes room for a mechanism of energy generation that is not exactly fusion, but rather a rearrangement of "clusters", referred to as photo-disintegration of Ne. In this process, the neon nuclei are broken by photons according to $^{20}\text{Ne} + \gamma \leftrightarrow {}^{16}\text{O} + \alpha$, and the α particles are soon captured in reactions of the type $^{20}\text{Ne} + \alpha \rightarrow {}^{24}\text{Mg} + \gamma$. If we look at the initial and final states, we can write effectively

$$\text{`` }^{20}\text{Ne} + {}^{20}\text{Ne} \rightarrow {}^{16}\text{O} + {}^{24}\text{Mg} \text{ ''}, \tag{4.42}$$

where the quotation marks remind us that this is *not* really the fusion of two neon nuclei. Shortly afterwards, for densities $\rho > 5 \times 10^6 \text{ g cm}^{-3}$ and $T \sim 2 \times 10^9$ K, oxygen can ignite, with an initial reaction

$$^{16}\text{O} + {}^{16}\text{O} \rightarrow {}^{32}\text{S}^*, \tag{4.43}$$

and analogously to (4.41), the excited sulphur nucleus decays to a number of possible final states whose integrated reaction rate is proportional to T^{35}. Very soon after that, another photodisintegration reaction (and the last) takes place from ^{28}Si:

$$\text{`` }^{28}\text{Si} + {}^{28}\text{Si} \rightarrow {}^{56}\text{Fe} + \gamma \text{ ''}, \tag{4.44}$$

where once again ^{28}Si denotes here a set of clusters with that mass number, but which is far from constituting a true silicon nucleus. Similarly, ^{56}Fe is a way of writing a series of elements with that mass number which immediately capture α particles and decay to form a distribution known as the *peak* elements of the iron group. What happens afterwards will be the subject of further study, in the collapse stage and the subsequent explosion [6].

The result of the whole sequence of reactions in the star structure leads to the *onion structure* shown in Fig. 4.22. It is important to understand that the time in which the star is sustained by each of the cycles gets rapidly shorter due once again to the vigorous expenditure of the available energy. These times are listed in Table 4.1 for a star of $20M_\odot$. The survival of the star depends on the fuel reservoir that remains in each case, but gets shorter and shorter, culminating in the supernova events to be described in the following Chapter.

In summary, the high-mass stars follow quite different trajectories from those of the low-mass stars in the HR diagram, as consequence of some important physical facts: initial energy production by the CNO cycle, convective cores, and temperatures/densities high enough to continue using the product nuclei for energy production. Some facts not discussed here, such as substantial mass loss, cause the HR diagram to shift to the left, and yellow or blue star explosions, instead of the red supergiants in the lower range up to about $15M_\odot$ (Fig. 4.23). We will see later how, in some well studied cases (the SN 1987A, for example), there is evidence for this and other complex characteristics that are currently under review.

An important insight into the stellar Physics of the most massive stars can be achieved by simultaneously imposing the Virial theorem and hydrostatic equilibrium, which determine the existence of solutions for the simplest stellar models. Using these concepts we can show how it is possible to characterize the end of the stellar sequence by observing the behavior of the gas and radiation pressures, and we shall see that the extracted value coincides, within uncertainties, with the maximum mass directly inferred for actual stars.

As a starting point, the hydrostatic equilibrium equation (4.5) states physically how gravity and pressure work against each other, showing that the long-lasting

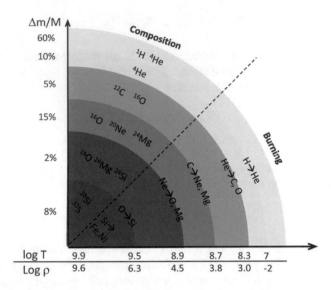

Fig. 4.22 Onion structure of a massive star that has already ignited all possible cycles to maintain itself and that has developed a core of "Fe" at the center. From [7]

Table 4.1 The different cycles of thermonuclear burning, together with the main products, ignition temperatures, and lifetimes for a star of $20 M_\odot$

Fuel	Main product	Secondary products	Temperature [10^9 K]	Duration of the cycle [year]
H	He	^{14}Ni	0.02	2×10^7
He	C, O	^{18}O, ^{22}Ne	0.2	10^6
C	Ne, Mg	Na	0.8	10^3
Ne	O, Mg	Al, P	1.5	3
O	Si, S	Cl, Ar, K, Ca	2.0	0.8
Si	Fe	Ti, V, Cr, Ni, Mn, Co	3.5	< 1 week

equilibrium solutions we call stars stem from an exact balance between the two. To make the situation even more transparent, we can define a "gravitational pressure" (a purely formal quantity) using the very general definition of pressure as the derivative of energy with respect to volume, viz.,

$$P_G = -\frac{\partial E_G}{\partial V} ,$$

and writing the gravitational energy in the form

$$E_G = -\frac{3}{5} \left(\frac{4\pi}{3} \right)^{1/3} \frac{G M^2}{V^{1/3}} .$$

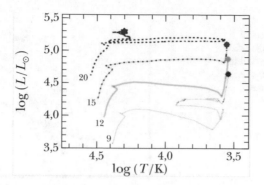

Fig. 4.23 Trajectories in the HR diagram for (non-rotating) stars of different masses and solar composition [8]. Note that due to the convective core the Schoenberg–Chandrasekhar limit is irrelevant and there is no ascent to the giant branch or helium flash. The mass loss is substantial and produces a return to blue for higher-mass stars (around $18M_\odot$ and beyond), until the moment of collapse and explosion (Chap. 5), indicated by stars in the diagram for each case. Credit: J.H. Groh et al., Astron. Astrophys. **558**, A131 (2013), reproduced with permission © ESO

This procedure will be used again to address the question of the existence of neutron stars (see Sect. 6.3.3).

The other fundamental ingredient is the Virial relation $\Omega_G + 2U = 0$. To tell us about the *end* of the high mass stellar solutions, we use the latter together with the hydrostatic equilibrium equation written in the form

$$\Sigma P_i = P_G , \quad \text{where} \quad P_G = CM^{2/3}\rho^{4/3} , \quad C = \frac{G}{5}\left(\frac{4\pi}{3}\right)^{1/3} ,$$

obtained by taking the derivative of (6.36). To show this, we recognize that the two physical ingredients in the pressure are the gas pressure and an increasingly important radiation pressure $P_{rad} \propto T^4$, that is $P_{tot} = P_{gas} + P_{rad}$. When we consider increasingly high star masses, the radiation pressure is initially small and can be neglected. On the other hand, the temperature gradient of the star grows to the point at which a convective adiabatic interior is achieved. In fact, at around $20M_\odot$, detailed calculations show that the convective structure occupies almost the whole star.

The problem of convective adiabatic equilibrium was addressed by Lord Kelvin a century ago. His reasoning was that the trajectories of stellar gas elements must at any point obey the relation $P = K\rho^{5/3}$, and therefore, since the fluid must be in equilibrium at all points inside, this mimics a polytropic relation of index 3/2, i.e., it has the same functional form of the latter. The threshold of the adiabatic convective condition means that the gradient reaches the adiabatic value

$$\nabla_{ad} = \frac{P}{T}\frac{dT}{dP} \equiv \frac{\partial \ln T}{\partial \ln P} \tag{4.45}$$

Fig. 4.24 Graphic
expression of hydrostatic
equilibrium. Solutions exist
as long as $P_{tot} = P_G$, but this
becomes impossible when
P_{rad} dominates the sum

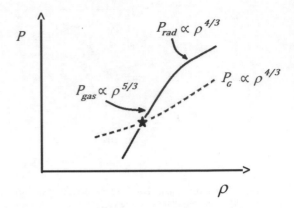

Thus for a constant value of the exponent, we can integrate (4.45) with the result

$$P_{gas} \propto T^{1/\nabla_{ad}} , \tag{4.46}$$

or equivalently, since $T \propto P/\rho$ for an ideal gas,

$$P_{gas} = K\rho^{1/(1-\nabla_{ad})} \equiv K\rho^{5/3} . \tag{4.47}$$

In this way we see that the adiabatic convective model due to Jeans is supported by a gas component that satisfies $P_{gas} = K\rho^{5/3}$. The existence of stellar models is guaranteed as long as $P_{tot} \approx P_{gas}$ because the slopes of P_{tot} and "P_G" $\propto \rho^{4/3}$ are different, and $P_{tot} = P_G$ can be satisfied (Fig. 4.24). However, when the mass grows, the radiation pressure grows as T^4, which is very fast, and the slope of the total pressure P_{tot} bends towards a slope of 4/3. The curves of P_{tot} and P_G become parallel and solutions (stars) cease to exist. This is expected to happen somewhere in the range 100–200$M_\odot$ and shows exactly how radiation pressure destabilizes the stellar structure when it dominates the total pressure.

As a corollary we believe that there is a maximum stellar mass determined by this condition, and empirical evidence has been sought to confirm this. Figure 4.25 shows some of the highest reported values, which do indeed confirm the general idea of there being an end to the stellar sequence. Accurate mass determinations becomes difficult for the highest values, since the photosphere is not well-located in wind environments. However, it is fairly safe to say that there is a simple reason for the maximum mass a star can have, even if its exact value remains somewhat uncertain.

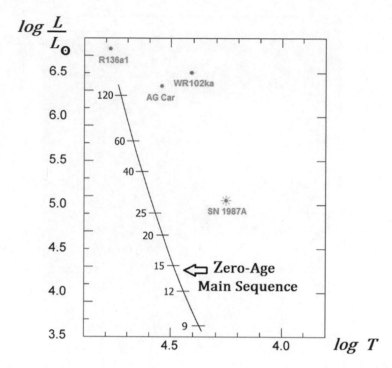

Fig. 4.25 Some of the largest stellar masses determined today. R136a1 was at first believed to be a single object of mass around $1000 M_\odot$, but is now known to be a triple system with a component of $320 M_\odot$. AG Car with mass around $100 M_\odot$ and WR102ka with mass $150 M_\odot$ are other examples of very massive stars. The progenitor of SN1987A with an initial estimated mass of about $19 M_\odot$ is also shown

References

1. S. Gregory, M. Zeilik, *Introductory Astronomy and Astrophysics* (Ed. Brooks Cole, London, 1997)
2. E. Bohm-Vitense, *Introduction to Stellar Astrophysics: Vol. 1: Basic Stellar Observations and Data* (Cambridge University Press, Cambridge UK, 1989)
3. C.A. Bertulani, P. Danielewicz, *Introduction to Nuclear Reactions* (CRC Press, Boca Raton USA, 2004)
4. D. Ostlie, B.W. Carroll, *An Introduction to Modern Stellar Astrophysics* (Pearson, New York, 1995)
5. G. Marx, Life in the nuclear valley. Phys. Edu. **36**, 375 (2001)
6. D. Clayton, *Principles of Stellar Evolution and Nucleosynthesis* (University Chicago Press, Chicago, 1984)
7. M. Spurio, *Probes of Multimessenger Astrophysics. Charged Cosmic Rays, Neutrinos, γ-rays and Gravitational Waves* (Springer, Berlin, 2018)
8. J.H. Groh et al., Fundamental properties of core-collapse supernova and GRB progenitors: predicting the look of massive stars before death. Astron. Astrophys. **558**, A131 (2013)

Chapter 5
Supernovae

5.1 Supernova Types and Classification

The visual recognition of the first supernovae (now called "historical") in the West dates back to the Middle Ages. Previously, supernovae that exploded in the years 185 A.D. and 393 A.D. had been visible, but were not recorded in the chronicles of the time. There is no clear evidence of these events in the West, although contemporary research has associated some accounts of religious authors with these events. The 1006 A.D. supernova, well documented by Chinese astronomers, would have caused bewilderment in Medieval Europe, then dominated by the Aristotelian dogma of the immutability of the heavens, but the event was not clearly recorded. A few decades later, the 1054 A.D. supernova (which today we know gave rise to the Crab Nebula) was recorded and studied in both East and West. We review the various supernova types, their basic Physics and related issues. Finally, and largely due to the scientific revolution, the supernovae observed and studied by Tycho (1572 A.D., Fig. 5.1) and Kepler (1604 A.D.) were widely discussed and the first steps were taken toward an understanding of this phenomenon [1], a task which is still ongoing, within a good overall picture, as we shall see in a moment.

In search of the physical origin of these events, the modern pioneers were the astronomers W. Baade and F. Zwicky. In a 1934 article [2] they realized that there were very large differences between the observed energy scales of the "novas" and the "super-novas", as they called them in these original writings, until then included in the same group. Baade and Zwicky were the authors of the first classification proposal, based on the presence or absence of hydrogen lines. According to them, the absence of hydrogen in the spectrum was a sign of an evolved star, possibly from the so-called population II (old), while if hydrogen was present, this could imply a population I star (young). Later it became clear that some "type I" events belonged to the young disk population, even though the energy scale was very similar in both cases (around 10^{51} erg). This forced a refinement of the classification. Soon, more astrophysicists built models that had as protagonists a massive white dwarf in a binary system (type Ia), or a massive star collapse, but where also He or Si could be

J. E. Horvath, *High-Energy Astrophysics*, Undergraduate Lecture Notes in Physics, https://doi.org/10.1007/978-3-030-92159-0_5

Fig. 5.1 *Left*: Tycho Brahe's original notes, recording the evolution of the observed brightness of the supernova in 1572 A.D., identified by the *arrow* at the top of the image. *Right*: False color mosaic image of the Tycho remnant in X-ray (Chandra) and infrared (Spitzer), showing the different chemical elements synthesized. It is now believed that the event was type Ia, of thermonuclear origin (see text). Credit: X-ray NASA/CXC/SAO, infrared NASA/JPL-Caltech, optical MPIA, Calar Alto, O. Krause et al.

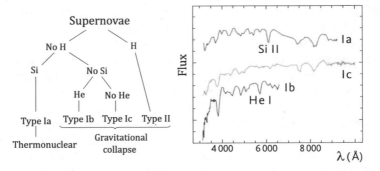

Fig. 5.2 Supernova classification according to the presence or absence of lines in the spectra (*left*). Envelope loss in the pre-supernova phases and the binary nature of the system are supposed to cause massive stars to produce type I supernovae, although in fact these also correspond to gravitational collapses. *On the right*, three spectra representing types Ia, Ic, and Ib from top to bottom, indicating some important lines. Credit: Swinburne University of Technology

absent for evolutionary reasons (called types Ib and Ic, Fig. 5.2), or where the whole envelope had to be ejected, i.e., with a massive presence of hydrogen (these were eventually called type II).

A number of studies conducted throughout the 20th century have shown that the explosion rates differ in different kinds of galaxy. For example, SNII are rarely observed in elliptical galaxies, which is thought to be due to the fact that these galaxies have no substantial stellar formation and contain few young stars. We can define 1 SNu (arbitrary unit) as the number of supernovae observed per century and per $10^{10} L_\odot$. We could then obtain the total rate in the Milky Way by simply multiplying the SNu observed in similar galaxies (those of Sb type) and noting that the Milky

Table 5.1 Chronology of historical supernovae in the last millennium

Name	Epoch	Distance [kpc]	Longitude	Latitude	Type
Lupus (SN1006)	1006	2.2	327.57	14.57	Ia
Crab	1054	2.0	184.55	−5.79	II (?)
3C58 (SN1181)	1181	2.6	130.73	3.07	II
Tycho	1572	2.4	120.09	1.42	Ia
Kepler	1604	4.2	4.53	6.82	(?) (disputed)
Cas A (no record)	1680	2.9	111.73	−2.13	Ib

Way has a luminosity $L_{MW} = 2 \times 10^{10} L_\odot$. This gives a rate of $1.3 \pm 0.9/second$ (including all types). However, the last observed supernova was Kepler's in 1604, more than 400 years ago, and we should have detected several other events since then. The relative rates, however, seem to be something like 20% type Ia, 70% type II and 10% type Ib/c [23]. Table 5.1 presents the historical supernovae in our galaxy, although some studies have obtained discrepant rates in which Type Ia dominate.

The theory behind type Ia and type II explosions is very complex, and to make matters worse, it is not enough to observe supernovae in distant galaxies to confirm them, since most predictions would require data from much closer events. However, as shown in Table 5.1, no observable event has occurred in the galaxy for nearly 4 centuries. Thus, the only well observed "historical" supernova to date took place in a nearby galaxy (SN1987A in the Large Magellanic Cloud), but it was enough to bring to reality most of the predictions of the collapse models that we will describe below.

5.2 Supernovae and Gravitational Collapse (Types II, Ib, and Ic)

The presence of hydrogen in type II supernovae, indicative of a young population, and the association of some events with the spiral arms of galaxies led to the idea that the progenitors of such events might by high-mass stars reaching the ends of their lifetimes. It later became clear that these stars would reach an extreme configuration with successive layers of increasingly heavy elements, as shown in Fig. 4.22. Now, accepting this evolution, it was important to ask how the star could explode. The description of gravitational collapse took a few decades, since it had to include some physical phenomena that were not understood in the first half of the 20th century, and others that still cannot be understood in a completely satisfactory way today. There is, however, an agreement about the main features and the sequence of events that we can address generically, as we shall do next.

The physical focus of collapse is on the behavior of the "Fe" core, where the quotes serve as a reminder that a variety of nuclides with a mass number close to 56 are included. As we have seen previously (Fig. 4.6), the binding energy is maximum near this value. The production of the "Fe" core must end the possible fusion reactions, since it is impossible to obtain fusion energy by fusing the elements of the iron peak. The growth of the core mass also has an absolute limit: the source of its pressure is degenerate electrons, which means that it can only grow to a value known as the Chandrasekhar mass. Without considering Coulomb, relativistic, and finite temperature corrections (the latter are quite important), this is given by

$$M_{\text{Ch}} = 1.46 \left(\frac{Y_e}{0.5} \right)^2 M_\odot .$$ (5.1)

Near this maximum, the central density and temperature are $\rho_c \sim 5 \times 10^9 \, \text{g cm}^{-3}$ and $T_c \sim 7 \times 10^9 \, \text{K} \approx 0.7 \, \text{MeV}$. The stability of the core is possible until it reaches this maximum. When this happens, two physical effects conspire to destabilize it. The first is that the density has increased so much that electrons are captured by the "Fe" nuclei in reactions of the type

$$e^- + A \rightarrow (A - 1) ,$$ (5.2)

that is, the number of particles sustaining the pressure decreases, so the pressure decreases. The second effect is that the thermal contribution of the pressure also decreases because energy is being used to break the "Fe" nuclei (photodisintegration) in the reaction

$$^{56}\text{Fe} \longleftrightarrow 13\alpha + 4\text{n} ,$$ (5.3)

an effect that goes in the same direction of decreasing the total pressure. As a consequence, the core collapses, the density increases, and this accelerates the e^- captures in an irreversible process [3].

When the density reaches values about 100 times higher than the initial one, a unique phenomenon in the contemporary Universe occurs in the collapsing core, related to the escape of neutrinos. In the early stages of collapse, neutrinos were able to escape unimpeded, taking away energy from the central region. But despite their having a very small cross-section, of the order of $10^{-44} \, \text{cm}^2$, more than 20 orders of magnitude smaller than an electron, for example, their trajectories will be affected by the increase in density. In fact, reactions of the type

$$\nu + \bar{\nu} \rightarrow e^+ + e^- , \quad \nu + A \rightarrow \nu + A ,$$ (5.4)

occur more frequently and cause the neutrino mean free path to decrease to the order of the radius of the collapsing core R. Numerically, the condition $n \times \sigma \times R \approx 1$ yields, for $R \approx 10$ km and the neutrino cross-section σ, a density $\rho_T \approx 4 \times 10^{11} \, \text{g cm}^{-3}$ above which neutrinos are retained in the collapsing core, i.e., they go from the free escape regime to the diffusive regime. This value is called the *trapping density*. Trapping causes the total number of electrons + neutrinos per baryon (Y_L) to remain *constant* at the initial numerical value of $Y_L \approx 0.37$. There is no more energy flowing out of the core and the process continues adiabatically from the trapping point onwards.

Studies of the collapsing core agree that the (Newtonian) equation of motion admits *homologous* solutions, in which the speed of the material elements is proportional to the radius at which they are found. This behavior is observed in simulations. If we call $\alpha(t)$ the constant of proportionality, a function of time, we have

$$\frac{u}{r} = \frac{\dot{\alpha}}{\alpha} = \text{constant} . \tag{5.5}$$

But of course the whole core cannot be in this homologous regime: as we consider matter at greater distances from the center by increasing the r coordinate, the inward speed of the matter u increases, and at some point it must satisfy

$$u + c_s = 0 , \tag{5.6}$$

where c_s is the speed of sound in the collapsing matter—at this point, an appreciable fraction of the speed of light. This means that matter in the inner region can maintain causal communication and maintain homology, but in the outer regions this is not possible. The point where the condition (5.6) is satisfied is called the *sonic point*, illustrated in Fig. 5.3.

From these considerations we see that, when the matter in the outer core continues to fall as a result of the strong gravity, a shock is formed at the sonic point (not at the center!) which encloses a mass of about $0.6 M_\odot$. This interior core becomes very hard when the saturation density $\rho_0 \sim 2.7 \times 10^{14}\,\text{g cm}^{-3}$ is reached, which breaks

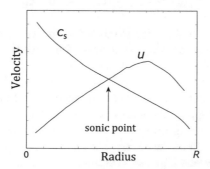

Fig. 5.3 The sonic point separates the collapsing core in an inner region ($R < R_s$) and an outer region ($R > R_s$). The homology relation can only be maintained in the inner region, while in the outer region the matter remains in almost free fall. This causal separation is important for the final destiny of the core

Fig. 5.4 *Left*: Analogy to visualize the effect of the sudden stiffening of matter in the inner core. The falling material forms a shock wave at the relevant "sonic point", similar to the reversal of the momentum of a particle that collides with a wall. *Right*: Shock wave calculated for a model of $15M_\odot$ with a core of mass about $1.5M_\odot$. The shock wave starts with positive velocity (*upper curve*) and advances towards the surface, but loses intensity rapidly and reverses itself after about 10–20 ms (*lower curves* with negative velocity) due to the energy losses discussed in the text

up all nuclear structure. It means that the matter falling onto it from above bounces when it strikes the edge of the inner core (Fig. 5.4), thus ending the homology.

Although no one doubts this sequence of events today, the most obvious expected outcome, namely, that the shock manages to eject the envelope and that this is the cause of the explosion, is *not* really what happens. As shown in Fig. 5.4 (right), the shock wave loses intensity as it travels outwards, mainly because the falling matter it meets on the way is still composed of nuclei: the outer core is no denser than ρ_0. The point is that dissociating these infalling nuclei costs the shock 1.8×10^{51} erg for every $0.1M_\odot$ crossed [the process described in (5.3)]. Since the simplest energy balance shows that the initial energy of the shock corresponds to the binding energy of the inner core (IC), i.e.,

$$E_{\text{shock}} \sim E_{\text{bind IC}} = 4\text{--}7 \times 10^{51} \text{ erg} , \qquad (5.7)$$

we have a paradoxical situation: the initial energy is more than enough to explode the star, but it is wasted in breaking the nuclei of the outer core, with mass $> 0.6M_\odot$ for almost any value of the progenitor mass.

After a series of studies that finally demonstrated the unfeasibility of the shock as a successful mechanism for the explosion, attention was focused on the fate of the core, now becoming a proto-neutron star, but at serious risk of collapsing to a black hole unless something happens soon after the shock stops. The key to this further evolution seems to be the energy that was released by compactification, and residing (because of the very high temperatures) in a sea of *neutrinos*, produced much more efficiently than photons under such conditions. A total of around 10^{53} erg is retained in the whole collapsed core, and it leaks away with difficulty as the condition $n \times \sigma \times R > 1$ is still satisfied, mainly because R is now quasi-stationary at 20–30 km, but the cross-sections of the neutrinos with matter grow a lot with temperature $T \sim 1$ MeV $\sim 10^{10}$ K. The neutrinos diffuse out of the proto-neutron

Fig. 5.5 The neutrinosphere in a just-formed proto-NS. The neutrinos are emitted from it with luminosity given by (5.7), and then some of them interact with the shock matter which "stopped" at the right of the figure

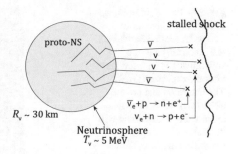

star on a diffusive timescale of about 1 s. Under the above conditions, the situation is analogous to the diffusion of photons in the solar interior, a regime that ends when they reach the optical depth surface $\tau < 1$, as described in Chap. 4. Thus, neutrinos are absorbed, re-emitted, and scattered many times until they finally reach the *neutrinosphere*, defined analogously to the photosphere, as if they were the Sun's photons. In the simplest hypothesis, the proto-NS emits like a black body, but for neutrinos, with a luminosity

$$L_v = \frac{7}{4}(4\pi R^2)\sigma T_v^4 \ \text{erg/s} , \tag{5.8}$$

a situation illustrated in Fig. 5.5. The factor $7/4$ stems from the fermionic character of the leaking neutrinos and expresses the difference with an ordinary black body.

Although the density of the material between the neutrinosphere and the stationary shock is much lower ($\leq 10^{13}$ g cm^{-3}, this region is sometimes called a *quasi-vacuum* precisely because of this sharp density difference), the rate of capture of the neutrino flux is not negligible. Because of the processes

$$\nu_e + n \rightarrow p + e^- , \tag{5.9}$$
$$\bar{\nu}_e + p \rightarrow n + e^+ , \tag{5.10}$$

there is a region of gain (before the shock) and loss (after the shock). The effective capture rate can be estimated as

$$\dot{Q} = F\sigma\langle E_v\rangle = \frac{L_v}{\langle E_v\rangle}\langle E_v^3\rangle , \tag{5.11}$$

and thus depends on the average energy of the neutrinos $\langle E_v\rangle$, the total luminosity L_v, and the detailed form of the spectrum (the average of the cube of the energy $\langle E_v^3\rangle$). These captures *do not* "push" the shock by transferring energy, but instead create conditions for it to expand again. This is why the mechanism is known as the *neutrino revival*. This would appear to be the cause of the expansion at the base of the envelope, with the explosion as its final outcome. We must insist that the impulse transfer to the shock by captured neutrinos is very small, but its dynamic

Fig. 5.6 Pre-existing image
of the progenitor star of
SN1987A (called Sanduleak
-69 202) (*left*) and the
supernova 1987A a few days
after the explosion (*right*).
The neutrinos that proved
SN 1987A to be a collapse
event were detected and will
be analyzed in Chap. 9.
Credit: Australian
Astronomical Optics (AAO)

effect, allowing a hydrodynamic wind solution, is possible (i.e., with finite pressure
at infinity) [4]. Core-collapse supernovae could quite properly be called *neutrino
bombs*.

In the last 400 years, the nearest supernova of this type was SN1987A, in the Large
Magellanic Cloud. This event is shown in Fig. 5.6. The supernova progenitor was
identified in pre-existing images, and had a mass of the order of 18–19$M_\odot$ according
to its position in the HR diagram. A complete reconstruction of its evolutionary
history suggests that the progenitor lived about 11 Myr, left the MS about 700 000 yr
ago, turning into a red supergiant soon afterwards, with a radius about three times
the orbit of the Earth around the Sun, exhausted its helium, and then lit its carbon
about 10 000 yr ago (when humankind was just taking up agriculture), burnt neon
from 1971 until 1983, oxygen from 1983 until February 1987, silicon for about 10
days in 1987, and finally exploded on 23 February of that year. All these periods
are subject to some uncertainty, but the basics are believed to be correct. The event
allowed us to begin the observation of neutrinos as a new discipline, and the details
that were reconstructed will be presented in Chap. 9.

We must stress that, although we believe that the shock revives as indicated,
the influence of "new" Physics (e.g., deconfinement of quarks subject to very high
pressure in the center) cannot yet be discarded [5]. In any case, nature does not care
much about this, and continues to make high-mass stars explode, as happened in
SN1987A and other frequently recorded collapse events. The are strong indications
that hydrodynamical instabilities in 3D are crucial for the success of the explosions.

Finally, it should be pointed out that the processes described here are essentially
the same as those that operate in the case of type Ib (without hydrogen) and type Ic
(without hydrogen or helium) supernovae. That is, the explosion mechanism is the
same even though the envelopes of the progenitors have been diminished or even
eliminated by the binary partner or stellar winds that are commonly observed in
high-mass stars. We will see later that there are concrete cases of supernovae that
reveal very high mass loss, and light curves indicative of these phenomena. A second
important issue is that there is evidence to suggest that collapsing stars at the lower

end of the mass range do not develop an "iron" core. The reason is that in the 8–$10 M_\odot$ range, nuclear reactions cannot take the star beyond the formation of lighter cores, composed of O, Mg, and Ne in degenerate conditions. Thus, when they reach their corresponding Chandrasekhar mass, electronic capture causes them to collapse with a practically invariant mass of about $1.38 M_\odot$. Throughout this collapse, the presence of oxygen ignited at $T \sim 2 \times 10^9$ K results in events that look like a thermonuclear supernova (see below), but with the formation of neutron stars with a low and fixed mass (around $1.25 M_\odot$), which results from the loss of around 10% of the original $1.38 M_\odot$ by the radiated neutrinos. Since there are many stars in this mass range, there must be several examples in the samples we have. A number of papers have suggested that the explosion that gave rise to the *Crab pulsar* in 1054 A.D. was of this type, with little luminosity and the production of the homonym pulsar. As a corollary, a "peak" is expected in the neutron star mass distribution at $1.25 M_\odot$, so establishing its presence should provide important support for these ideas.

Finally, not all core-collapse explosions have given rise to a pulsar. There are hot sources (Central Compact Objects, or CCOs) in some of them which do not pulse, and this is attributed to a combination of low magnetic field and low rotation rate. The "classic" explanation that the pulsar beams may point away from us does not seem to match the statistics. In other words, the neutron stars "should be there", but are not really detected.

5.3 Thermonuclear Supernovae

In the pioneering studies of Baade and Zwicky in the 1930s, all "super-novas" belonged to the same class, although there was a difference that would become important for the refined classification: the presence or absence of hydrogen in the spectrum, as previously pointed out. As work progressed, it was observed that the occurrence of supernovae without hydrogen (and without silicon, corresponding to what we know as type Ia) pointed to an *old population*, not only because of the above-mentioned absence of hydrogen (characteristic of an evolved progenitor), but also because they were not, on average, located in the plane of the galaxies like the young stars (so-called Population I stars). Thus, they began to think about what kind of evolved progenitor (Population II stars) could suddenly release an enormous amount of energy (the roughly 10^{51} erg observed) and what process would allow this release.

The analysis of possible energy sources led them to consider the uncontrolled fusion of carbon into a white dwarf as the most viable mechanism that satisfied the given conditions. But still, there were (and there are) two scenarios for fusion of the carbon: the white dwarf could ignite the carbon by the long term accretion effect from a "normal" companion, or it could also happen in binary systems, in the final stage when two white dwarfs merge and matter compresses and heats up. The first scenario is described as *single-degenerate* and the second as *double-degenerate*, names that

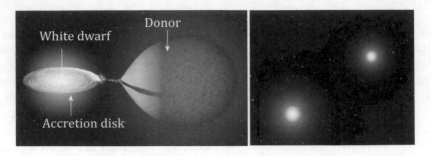

Fig. 5.7 The two possible scenarios that would allow carbon ignition in a white dwarf. *On the left*, the single-degenerate case, where a white dwarf accretes matter from a normal post-MS companion, and *on the right*, the double-degenerate case, where two white dwarfs end up merging after a long time, when their orbit decays. From [6]

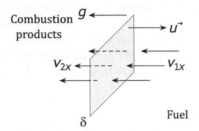

Fig. 5.8 Combustion front δ, supposedly locally flat and of small thickness compared with the dimensions of the system. Region 1 contains the fuel (in this case, carbon) and region 2 the fusion products. u is the speed of propagation of the front in the unburnt medium of region 1

correspond to the number of (degenerate) white dwarfs involved. Figure 5.7 illustrates these two scenarios.

Studies of carbon ignition in the single-degenerate case still present several uncertainties. It is a common assumption that a single white dwarf will not be able to reach these conditions unless it is near the Chandrasekhar limit. The ignition temperature for densities around 10^9 g cm^{-3} (appropriate for the center under these conditions) is above 5×10^8 K. Since oxygen has an ignition temperature about five times higher, only the carbon component of the white dwarf can fuse under these conditions (see Chap. 6). Once the carbon is ignited, the combustion front should spread more or less spherically to the surface. This type of phenomenon can be described by considering the conservation of physical quantities, as we will see below (Fig. 5.8).

To formulate the problem, we can write down the equations of conserved mass flux, energy, and momentum through the surface δ with the result [7]

$$\rho_1 v_{1x} = \rho_2 v_{2x} \equiv j \, , \tag{5.12}$$

$$\frac{1}{2} v_{1x}^2 + \omega_1 = \frac{1}{2} v_{2x}^2 + \omega_2 \, , \tag{5.13}$$

$$P_1 + \rho_1 v_{1x}^2 = P_2 + \rho_2 v_{2x}^2 \, , \tag{5.14}$$

where $\omega = \epsilon + P/\rho$ is the enthalpy function per unit mass, v_{1x} and v_{2x} are the normal speeds at the surface, and P and ρ are the pressures and densities of the two fluids. Equations (5.12)–(5.14) are written in the system moving with the front δ in the Newtonian approach, and since they merely express conservation laws, they must be satisfied by *any* form of combustion.

Using the first two equations above and defining the volume $V = 1/n$, we immediately arrive at

$$j^2 = \frac{P_2 - P_1}{V_1 - V_2} \, , \tag{5.15}$$

which relates the mass flow to the quantities on each side of the front δ. Finally, substituting in (5.12)–(5.14) and arranging, we have

$$\omega_1 - \omega_2 + \frac{1}{2}(P_2 - P_1)(V_1 + V_2) = 0 \, , \tag{5.16}$$

known as the *Chapman–Jouguet (CJ) adiabat*. It differs from Hugoniot's adiabat—which describes a discontinuity without combustion and without chemical change of the gas, i.e., a shock—because the enthalpy function ω is different on each side. If the gas did not change state we would have $\omega_1 = \omega_2$. However, for combustions $\omega_1 \neq \omega_2$ because the fluid behind the front δ, in the burnt region 2, is in general a different function. The CJ adiabats are curves specified by two parameters: if we know the state of the gas before burning, specified by P_1, V_1, we can determine P_2, V_2 given the mass flux j (Fig. 5.9).

As the mass flux j is positive, we need either $P_2 > P_1$ and $V_1 > V_2$, or alternatively $P_2 < P_1$ and $V_1 < V_2$. The first solution leads to the upper branch of the CJ adiabat and the combustions are called *detonations*; the second possibility is achieved on

Fig. 5.9 Chapman–Jouguet adiabat. The gas (in the supernova case, carbon) burns in state P_1, ω_1 and the products need to be above the CJ curve to satisfy the conservation laws. Depending on the mass flux j, this combustion may end in the lower or upper branch of the adiabatic, two quite different combustion modes

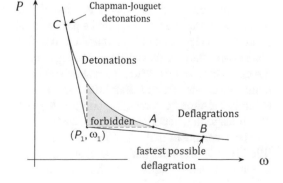

the lower (quasi-horizontal) branch, between points A and B in Fig. 5.9, where the combustions are called *deflagrations* (ordinary combustions). In the intermediate gray region the mass flux is imaginary, and therefore these solutions do not exist in nature.

The question is now: what is the mode of combustion in type Ia supernovae? Physically, the two paths are quite different. Ordinary deflagrations happen when the heat released in the reaction zone inside the front δ diffuses and helps burn the fuel ahead (think, for example, of setting fire to a sheet of paper). The basic quantity that determines the spreading is the thermometric conductivity, given by $\chi = \kappa_{\mathrm{th}}/C_P\rho$, where κ_{th} is the thermal conductivity and C_P is the heat capacity at constant pressure. Using this, we can estimate the width of the flame by $\delta = (\chi\tau)^{1/2}$, where τ is the characteristic reaction time and the speed is of the order of $u_{\mathrm{def}} \sim \delta/\tau \approx (\chi/\tau)^{1/2}$. Although the latter is slow in everyday life (again, think of the sheet of paper), it can be very fast, but will always be subsonic in a white dwarf. In contrast, the detonations of the upper branch of Fig. 5.9 are ultimately mediated by a shock which "burns" the particles behind it. The shocks are always supersonic in region 1 ahead of the flame—think what happens when you are standing waiting for the subway, and a blast of wind suddenly hits you: this is the shock produced by the train when it travels through the tunnel; people standing at the platform are not "warned" because it is supersonic, and it arrives even before the mechanical sound of the train. Thus, the medium in region 1 *cannot expand* before the flame reaches it, and the combustion is always total, never intermediate or partial [6, 8].

The key to determining which mode occurs in a star lies once again in the observations. SNIa events have light curves compatible with the production (synthesis) of at least $0.6M_\odot$ of nickel, but also 0.2–$0.3M_\odot$ of Si, Ar, Ca, and S, i.e., elements of intermediate mass that are "partially burned ashes", as happens with embers that did not completely burn in a barbecue. Less common elements such as ^{54}Fe and ^{58}Ni that would spoil the observed nucleosynthesis cannot be produced in any quantity, since their actual observed abundance is very small.

Since detonations take place in conditions of *nuclear statistical equilibrium*, where the number of protons and neutrons does not change in the reactions (there is no time for this), and only when this equilibrium is violated can combustion leave intermediate elements as a result, we deduce that there must be at least one stage of the combustion that is in the deflagration mode. Thus, the propagating front can "warn" the matter ahead (since it is subsonic). The front therefore expands because of the waves that travel ahead, and a partial combustion is possible. But on the other hand, there is the need to have abundant nickel, easily produced by detonations. This gives rise to the idea that the combustions may begin as deflagrations, then "jump" to the detonation branch when the instabilities of the flame deform it and increase the combustion rate. These instabilities can be seen in any film where the protagonists explode gasoline or something similar: the calculation of the front (using Euler's equations with reaction terms) is shown in Fig. 5.10 for a thermonuclear supernova.

The mechanism that transforms a deflagration into detonation is not well identified in this problem, but these changes have already been observed and studied in laboratories. In supernovae, this model is called the *deflagration-to-detonation*

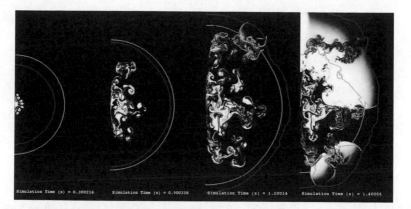

Fig. 5.10 Calculation showing the instability action (Landau–Darrieus, Rayleigh–Taylor, and others) in front of carbon combustion at four different moments: 0.3 s, 0.9 s, 1.2 s, and 1.4 s after the beginning of the combustion in the center of the star, on the left [9]. Credit: Alan Calder and Dean Townsley

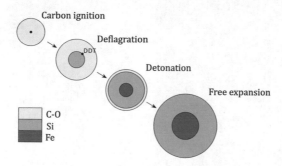

Fig. 5.11 DDT model. At the top of the figure, the white dwarf lights carbon and the front spreads like a deflagration until the instabilities force it to "jump" to the detonation branch. Initially, intermediate mass elements are produced, followed by a large amount of nickel [10]

transition (DDT). It is compatible with observations, but is not consensual among researchers in this area (Fig. 5.11).

In any case, there is ample evidence to show that type Ia supernovae release much more energy than is necessary to unbind a white dwarf: these explosions never leave a stellar remnant (unless it is a single-degenerate explosion where the white dwarf is accreting helium from its companion, in which case there may be a partial remnant known as a *zombie*). There are hundreds of light curves available and in all of them a stage can be identified in which, after the maximum, the temporal decay of the curve coincides with the half-life of $^{56}Ni \rightarrow {}^{56}Co$ (6.1 d), followed by another stage where the decay $^{56}Co \rightarrow {}^{56}Fe$ (78.8 d) is clearly visible. This is why we insist on the abundant production of ^{56}Ni as a necessary requirement. This sequence after the maximum, which is determined by the condition of the photons' diffusion time in

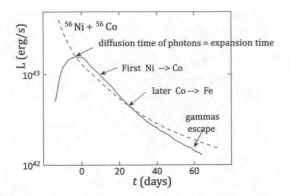

Fig. 5.12 Light curve of a SNIa with the main characteristics that determine its shape identified explicitly. Without energy sources, the temporal decay would be very fast and incompatible with observations [8]

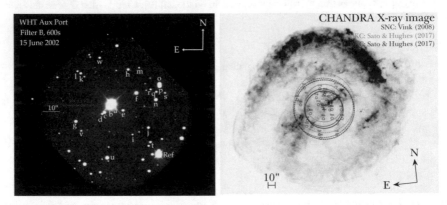

Fig. 5.13 *Left*: Search for the surviving companion star [11]. *Right*: The Kepler remnant (still disputed as an SNIa explosion). No candidates have been identified within the circles marking the central region of the explosion [12]. © AAS. Reproduced with permission. Credit: P. Ruiz-Lapuente

the expanding envelope, equals the time scale of the expansion, and the subsequent escape of the gamma photons from the remnant are indicated in Fig. 5.12.

One last point worth highlighting is the series of recent attempts to directly determine whether supernovae correspond to the single-degenerate or double-degenerate scenario. There are some ways to investigate this possibility by observing the historical type Ia supernova regions for which there is no doubt as to their nature (SN1006 and Tycho, see Table 5.1). The simplest way is to look, close to the center of the explosion, for some star altered by the shock's passage. This star should be partially swept, and if so, suspected to be the one that would transfer mass until the carbon ignited (Fig. 5.7 left). The search in Tycho's remnant only revealed one candidate, possibly a halo star, unrelated to the explosion region that is there by chance (background). In the Kepler remnant no candidate was found (Fig. 5.13). Thus we could conclude that there is no direct evidence of the single-degenerate scenario and that the explosions must have been produced by a merging of two white dwarfs. However, analysis of a third remnant, 3C 397, from some 2000 years ago, by the Suzaku satellite showed that this remnant must have been produced by the explosion of a single white dwarf. This is due to the observation of the amount of nickel, magnesium, iron, and chromium and comparison of the models, which are quite different in the two cases and which excludes a binary system of two white dwarfs as progenitor for 3C 397. Thus emerges the idea that the two scenarios could produce SNIa, although most of them were due to WD–WD binaries.

5.4 Type Ia Supernovae and Cosmology

The fact that supernovae reach negative absolute magnitudes indicates that, in the hypothesis of being able to find a standardization for the light curve, they would be excellent "rulers" for measuring very large distances, since they are seen up to cosmological scales. This observation is the basis of the work that sought to systematize the observations and convert supernovae into tools for Cosmology. For example, with a large telescope that can detect a visual magnitude around 25 it would be possible to see a supernova up to $z \sim 3$, that is, whose light was emitted when the Universe was just a quarter of its current scale. The record belongs to the detection of a supernova that exploded when the Universe was less than 3 Gyr old [14].

If we collect the light curves of type Ia supernovae, identified as such by their spectra, there is a variety that seems indisputable (Fig. 5.14). But since they are cosmological sources, we must apply the corrections of the cosmological model to correct both the observed duration and luminosity. When this procedure is carried out, the curves converge to a universal form (Fig. 5.14), and astrophysicists speak of a *calibration* (known as the Hamuy–Phillips calibration, [15]). The result is that the maximum luminosity is only a function of the temporal width: the wider the curve, the brighter the object. This is interpreted as showing the universality of the

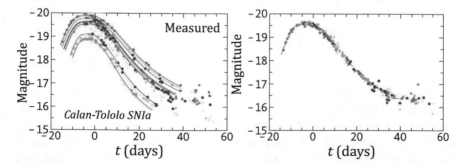

Fig. 5.14 Light curves of several SNIa for different redshifts (*left*), and the shape obtained after applying the corrections (*right*). This procedure is known as the Hamuy–Phillips calibration. Credit: Calan–Tololo SNIa Survey

Fig. 5.15 Type Ia
supernovae in the
corresponding Hubble
diagram. Although visually
not obvious, the statistical
evidence for a deviation in
the direction of the blue
curve model, where the rate
of expansion accelerates, is
strong. The other two cases
with decelerated expansion
are strongly disfavored [13].
Copyright (1999) National
Academy of Sciences, USA

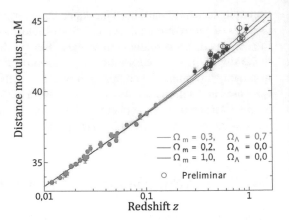

light curves of explosions: all differences are due to Cosmology. Thus, if we observe
more distant supernovae, we can know what they are like intrinsically (by applying
the calibration) and with it test the cosmological model. Supernovae thus became
"standard candles" for Cosmology.

It was precisely the implementation of these ideas that led two independent
teams—the *Supernova Cosmology Project* and the *High-Z Supernova Search Team*—
to announce that the data favored a model in which the expansion of the Universe is
accelerating. The reasoning is easy to follow: the supernova data "calibrated" with
the correction of the "standard" cosmological model used until then show that the
supernovae are systematically *fainter* for greater distances. Thus, the simplest solution is to think that the Universe has expanded faster and dragged supernovae to
greater distances. If the model of the Universe does not have this extra expansion,
supernovae cannot be placed in it (Fig. 5.15).

Note that so far we have not questioned the *cause* of the acceleration, which is
a separate problem. It has only been stated that the expectation of the magnitudes
of the SNIa according to the decelerated models does not match the observed data.
For an everyday analogy, we can imagine that we have identical lanterns and several
carriers at different distances: by measuring the flux and knowing how much they
emit intrinsically, we can calculate the distances for each one. This is exactly what
the research teams did with the measurements announced some 20 years ago, and
which are even firmer now that a much larger and better studied sample has become
available.

To end this discussion, we know that the simplest cosmological hypothesis to
explain this acceleration is to introduce a *cosmological constant* Λ that acts dynami-
cally as a repulsion imposing a positive acceleration on the Universe. But other possibilities are being considered, for example, modified gravitation and extra dimensions,
among the best known. There have also been attempts to challenge the interpretation
of the observations, suggesting that there could be an extinction of the magnitudes
over such distances. This is a very complicated hypothesis because the known extinction always depends on the wavelength, while the measurements indicate that it is

universal (independent of λ) in the case of SNIa (one of the teams made this working hypothesis public at the time of the announcement).

Finally, we should point out that thermonuclear supernova modeling now introduces a somewhat worrying element into the discussion: if the two scenarios (single-degenerate and double-degenerate) can produce events, and furthermore WD–WD binaries are the majority, why would the light curves be identical? Would a large dispersion be expected between them, visible in the calibration on the right of Fig. 5.14 after the cosmological correction has been applied? Much theoretical and observational work will be required to answer this question.

5.5 Superluminous Supernovae

The recognition of the existence of supernovae with energies much higher than the "standard" value of around 10^{51} erg, and even higher than the "hypernovae" (type Ic supernovae) that reach up to 10^{52} erg, dates back to the first decade of the 21st century [16]. Until that moment, the classification already discussed worked very well, with the exception of having to include SNIc with wide lines, associated with the occurrence of a gamma-ray burst (see Chap. 11). But progenitors of mass greater than $10 M_\odot$ seemed to explode according to the given classification, while the lighter ones were supposed to lead to the electron-capture events mentioned at the end of Sect. 5.2, and there were no major problems on the horizon, at least from this point of view (Fig. 5.16).

However, it became clear that some events with "anomalous" light curves should correspond to explosion energies of up to 10^{53} erg, without it being clear how this happened. These were referred to as *superluminous supernovae* (SLSN), defined empirically as ones where the absolute optical magnitude is less than -21, i.e., more luminous, as indicated by their more negative magnitude. We will soon examine our present theoretical understanding of these events.

There are three basic models in the literature for SLSN explosions. Pair instability supernovae, the collision of ejection with material in the circumstellar medium, and the injection of energy by a magnetar—a highly magnetized neutron star, discussed in Chap. 6. We will give a brief description of each of them below.

Pair instability is an expected phenomenon in the extreme circumstances present in very high mass stars. Motivated by studies of stars with masses of several hundred solar masses at very low metallicity (called population III by astronomers), possibly associated with the first star generations when the first structures formed in the Universe, there was interest in understanding their evolution and corresponding nucleosynthesis. At the most advanced stage, a star with mass in the range 80–$100 M_\odot$ or above with a helium core can convert photon energy into pairs e^+ and e^-, and taking into account that radiation makes a substantial contribution to the total pressure, this can cause the star to collapse. Throughout the collapse, there is explosive fusion of carbon and oxygen and the release of energy is more than enough to unbind the star, producing an enormous amount of ^{56}Ni. The observations of some SLSN require at least $10M_\odot$ of nickel, a result that seems possible for "zero metallicity" stars between $150M_\odot$ and $250M_\odot$ [19]. It could also happen for higher metallicities if the mass loss is suppressed, for example, by high magnetic fields. A variant of this scenario is that, depending on the mass, pair instability can lead to mass ejection pulses of around $10M_\odot$, whence one would observe a supernova that lasts several years rather than exploding in one go. Supernovae inside planetary nebulas (SNIPs) have been associated with the "slow decay" events of the SLSN Ic light curves in Fig. 5.17.

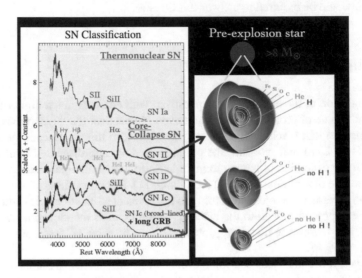

Fig. 5.16 Spectra of collapse events (*lower part*) and thermonuclear supernovae (*upper part*), showing the structure of the parents (*right*). Electron-capture events do not appear in this figure [17]. Credit: M. Modjaz

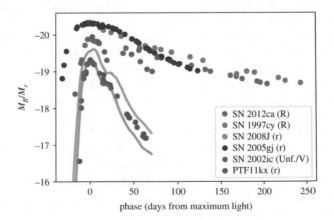

Fig. 5.17 Light curves of several SLSN. The curves decay over a few months or even longer, in some cases even of order 1 yr [18]. Credit: M. Fraser

Note that it is almost impossible for these progenitors to form in a high-metallicity environment. It is not clear whether the rapidly decaying SLSN Ic requires another model, and the calculated spectra are not in full agreement with observations, but at some point the identification of pair instability SNs must be achieved.

The case of the collision of the ejected gas with previously existing matter in the circumstellar environment is paradigmatic in Astrophysics. These collisions lead to shocks, and these are a very effective way of converting kinetic energy into thermal energy, later radiated away. Numerical studies show that, if the circumstellar material is dense, ejections of over $10 M_\odot$ bear similarities, for example, with the light curve of SN2006gj (red points in Fig. 5.17). A pulsed pair instability mechanism also leads to consider it as a candidate, since at the bottom this populates the circumstellar medium with successively ejected shells. The SLSN Ic could also be produced this way, but this depends a lot on the composition of the ejected material.

Finally, the scenario where there is a (huge) injection of energy by a newborn magnetar that acts as a kind of piston was studied by Allen and Horvath [20] and later applied to the production of SLSN by Kasen and Bildsten [21] and Woosley [19]. Basically, the energy of a fast rotating magnetic dipole injected into the remnant at a time t is

$$
E_{\text{inj}} = \frac{L_0}{\tau_0^{-1} + t^{-1}}
$$

$$
= 2 \times 10^{52} \left(\frac{1 \text{ ms}}{P_0} \right)^2 \frac{1}{1 + 1.6 \times 10^{-3} \left(\frac{10^{14} \text{ G}}{B} \right)^2 \left(\frac{P_0}{1 \text{ ms}} \right)^2 \left(\frac{1 \text{ yr}}{t} \right)} \text{ erg} ,
$$

$$(5.17)$$

where P_0 is the initial period (of the order of 1 ms, otherwise the magnetic field B could not grow enough) and

$$
\tau_0 = 0.6 \text{ d} \left(\frac{10^{14} \text{ G}}{B} \right)^2 \left(\frac{P_0}{1 \text{ ms}} \right)^2
$$

Fig. 5.18 Comparison between some slow-growing SLSN (SN 2011kl) and one of extreme brightness (ASASSN-15lh) with the magnetar model that best reproduces them (*full red curves*). From [22] © AAS. Reproduced with permission

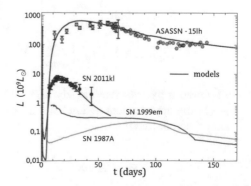

is an injection timescale. We see that, if the field grows, there will be enough injected energy to explain the light curves. A more detailed comparison is shown in Fig. 5.18.

5.6   Expansion of Supernova Remnants in the Interstellar Medium

In a first approximation, all supernovae are essentially point explosions that later expand in the surrounding interstellar medium (ISM), regardless of the specific type. We can thus formulate a general description of the expansion evolution, provided that some simplifications are admitted. One of these hypotheses is that the density of the interstellar medium can be described as a power law of the distance $\rho_{ISM} = \rho_0 r^{-k}$, which is quite realistic in cases, for example, of mass loss in the pre-supernova stages, and includes the case of $\rho_{ISM} = \rho_0 = $ constant, when $k = 0$. Thus, the total mass of the remaining M_{SNR} grows as the remnant sweeps the ISM according to the formula

$$M_{SNR} = M_{ej} + \int_0^{R_{SNR}} 4\pi r^2 \rho_{ISM} \, dr = M_{ej} + \frac{4\pi}{3-k} \rho_0 R_{SNR}^{3-k} , \qquad (5.18)$$

for $k < 3$. From (5.18) we see immediately that, when the remnant reaches a radius of approximately 1 pc, the swept mass is comparable to the mass ejected from the SN, denoted M_{ej}, almost independently of the value of k. Observations indicate, on the other hand, the injection of a "standard" energy from the explosion of $E_{exp} = 10^{51}$ erg, which will be our reference below. In the early stages of the explosion, $M_{SNR} \approx M_{ej}$, the energy losses of the remnant due to radiation are small (i.e., the expansion is essentially adiabatic), and the solution of the equations of motion in this phase of free expansion is simply

$$\dot{R}_{SNR} = \sqrt{2E_{exp}/M_{ej}} \equiv v_{SNR} , \qquad R_{SNR} = t\sqrt{2E_{exp}/M_{ej}} .$$

Scaling to typical values, these can be written:

$$R_{SNR} = 0.32 \left(\frac{M_{ej}}{10M_\odot}\right)^{-1/2} \left(\frac{E_{exp}}{10^{51} \, erg}\right)^{1/2} \left(\frac{t}{100 \, yr}\right) pc , \qquad (5.19)$$

$$v_{SNR} = 3.2 \times 10^3 \left(\frac{M_{ej}}{10M_\odot}\right)^{-1/2} \left(\frac{E_{exp}}{10^{51} \, erg}\right)^{1/2} km \, s^{-1} . \qquad (5.20)$$

As time goes by, we have already said that the swept mass grows until it equals M_{ej} and the approximations used so far are no longer valid. The instant at which $M_{SNR} \approx M_{ej}$ occurs is

$$R_{SNR} = 4.8 \left(\frac{M_{ej}}{10M_\odot}\right)^{1/3} \left(\frac{n_{ISM}}{1 \, cm^{-3}}\right)^{-1/3} pc , \qquad (5.21)$$

$$t = 1.4 \times 10^3 \left(\frac{M_{ej}}{10M_\odot}\right)^{5/6} \left(\frac{n_{ISM}}{1\,cm^{-3}}\right)^{-1/3} \left(\frac{E_{exp}}{10^{51}\,erg}\right)^{-1/2} \quad yr\,, \qquad (5.22)$$

hence between 1 and 2 millennia after the explosion. In this new stage after free expansion, the internal energy of the gas increases at the expense of the kinetic energy of the SNR, and one should thus consider the internal pressure

$$P_{int} = (\gamma_{int} - 1)U_{int} \left(\frac{4\pi R_{SNR}^3}{3}\right)^{-1}\,,$$

where U_{int} is the internal energy, the difference between the E_{exp} and the kinetic energy $M_{SNR} v_{SNR}^2/2$ of the SNR. One usually makes the *thin shell approximation* here, assuming that the whole mass is concentrated in one thin shell, and assumes *strong shock conditions*, where the "density jump" across the shock is maximum, i.e., $\rho_{SNR} = 4\rho_{ISM}$ [7]. The equations of motion

$$\frac{d(Mv)}{dt} = 4\pi R^2 P_{int}\,, \qquad (5.23)$$

$$U_{int} = E - \frac{1}{2}Mv^2 + \frac{L_0}{\tau_0^{-1} + t^{-1}}\,, \qquad (5.24)$$

can be solved. The last term of (5.24) represents an injection of energy from the inside, for example, by a neutron star in a very fast and magnetized rotation, as we saw in the last section. Without this last term, the system can be solved by proposing a power law for the radius of the remnant, and then determining the exponent. These solutions were already obtained in the classical works of Sedov and Taylor in the 1950s [23]. In this Sedov–Taylor phase, the radius increases as

$$R_{SNR} = \left(\frac{50 E_{exp}}{9\pi \rho_{ISM}}\right)^{1/5} t^{2/5}\,.$$

Expressed in terms of typical scales of the problem, we have

$$R_{SNR} = 15 \left(\frac{E_{exp}}{10^{51}\,erg}\right)^{1/5} \left(\frac{n_{ISM}}{1\,cm^{-3}}\right)^{-1/5} \left(\frac{t}{10^4\,yr}\right)^{2/5} \quad pc\,, \qquad (5.25)$$

$$v_{SNR} = 580 \left(\frac{E_{exp}}{10^{51}\,erg}\right)^{1/5} \left(\frac{n_{ISM}}{1\,cm^{-3}}\right)^{-1/5} \left(\frac{t}{10^4\,yr}\right)^{-3/5} \quad km\,s^{-1}\,. \qquad (5.26)$$

Physically, we find that the thin shell decelerates *before* the gas that comes soon after, generating an *internal shock* that propagates inwards, called the *reverse shock* (Fig. 5.19). The reverse shock heats up the interior gas and this energy ends up

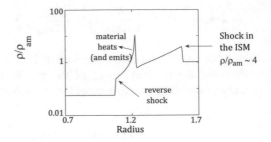

Fig. 5.19 Schematic picture of the shock propagating in the ISM (indicated) and the reverse shock generated by the dynamics of the thin inner gas shell

being radiated, observed as continuous emission in X-rays, with temperatures $T_{SNR} \geq$ 10^6 K. Due to these losses, the external shock on the ISM should at some point cease to be adiabatic, since the conditions around 2×10^4 yr favor the end of the S–T stage and the beginning of a new stage where the shell of the remnant cools down by emission due mainly to bound–free, free–free, and bound–bound transitions, which produce a loss term proportional to $T_{SNR}^{-1/2}$. For these reasons, it may be that remnants will be very difficult to observe far beyond this cooling age, since they become quickly "invisible" as cooling progresses.

The next stage, called the *snow-plow phase* by analogy with the situation where energy is spent to push a large swept mass, cannot even be described exactly. In practice, it is assumed that the thin shell becomes colder and denser, but that the gas swept inside by the reverse shock continues to expand adiabatically, and thus satisfies $P_{SNR} V_{SNR}^{5/3} = $ constant. The solutions to the motion equations are once again obtained by proposing a power law for R_{SNR} and using the previous adiabatic stage as initial condition, which results in

$$R_{SNR} = 30 \left(\frac{E_{exp}}{10^{51} \, \text{erg}} \right)^{11/49} \left(\frac{n_{ISM}}{1 \, \text{cm}^{-3}} \right)^{-13/49} \left(\frac{t}{10^4 \, \text{yr}} \right)^{2/7} \text{pc} , \qquad (5.27)$$

$$v_{SNR} = 120 \left(\frac{E_{exp}}{10^{51} \, \text{erg}} \right)^{-4/49} \left(\frac{n_{ISM}}{1 \, \text{cm}^{-3}} \right)^{-13/49} \left(\frac{t}{10^4 \, \text{yr}} \right)^{-5/7} \text{km s}^{-1} . (5.28)$$

Finally, after the snow-plow stage, which lasts about 10^6 yr, the shell breaks into fragments that mix with the ISM gas, which has an observed root mean square (rms) speed of about 20 km/s. Clearly the remnant ceases to be visible here, and as we have pointed out before, it is even possible that this disappearance happens long before this kinematic fusion. Almost all the remnants detected so far (more than 200) are in the S–T stage, although those that gave rise to strongly magnetized neutron stars (magnetars) may receive additional energy and appear to be older than they really are because of the last term of (5.24). Figure 5.20 presents a scheme of successive exponents in each phase of expansion that summarizes this discussion [24].

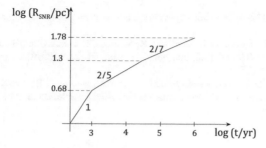

Fig. 5.20 Exponents of the supernova remnant evolution through the free expansion, Sedov–Taylor, and snow-plow stages. The approximate values for the transition radius between these stages are indicated on the vertical axis for ISM standard density $n_{ISM} = 1$ cm^{-3}

References

1. D.H. Clark, F.R. Stephenson, *The Historical Supernovae* (Pergamon, London, 1977)
2. W. Baade, F. Zwicky, On super-novae. Proc. Natl. Academy Sci. USA **20**, 254 (1934)
3. H.A. Bethe, *Supernova Theory* (World Scientific, Singapore, 1994)
4. J.W. Murphy, A. Burrows, Criteria for core-collapse supernova explosions by the neutrino mechanism. Astrophys. J. **688**, 1159 (2008)
5. O.G. Benvenuto, J.E. Horvath, Evidence for strange matter in supernovae? Phys. Rev. Lett. **63**, 716 (1989)
6. P. Hoeflich, Explosion Physics of thermonuclear supernovae and their signatures, in *Handbook of Supernovae*, eds. A.W. Alsabti and P. Murdin (Springer, Berlin, 2017), p. 1151
7. L.D. Landau, E.M. Lifshitz, *Fluid Mechanics* (Pergamon Press, Oxford, 2013)
8. D. Branch, J.C. Wheeler, *Supernova Explosions* (Springer, Berlin, 2017)
9. A.C. Calder, B.K. Krueger, A.P. Jackson, D.M. Townsley, The influence of chemical composition on models of Type Ia supernovae. Frontiers in Physics **8**(2), 168–188 (2013)
10. V. Gamezo, A. Khlokhov, E. Oran, Deflagrations and detonations in thermonuclear supernovae. Phys. Rev. Lett. **92**, 211102 (2004)
11. P. Ruiz-Lapuente et al., The binary progenitor of Tycho Brahe's 1572 supernova. Nature **431**, 1069 (2004)
12. P. Ruiz-Lapuente et al., No surviving companion in Kepler's supernova. Astrophys. J. **862**, 124 (2018). https://doi.org/10.3847/1538-4357/aac9c4
13. R.P. Kirshner, Supernovae, an accelerating Universe and the cosmological constant. Proc. Natl. Acad. Sci. **96**(8), 4224–4227 (1999)
14. M. Smith et al., Studying the ultraviolet spectrum of the first spectroscopically confirmed supernova at redshift two. Astrophys. J. **854**, 37 (2018)
15. M. Hamuy, *Low-z type Ia supernova calibration*, eds. A.W. Alsabti and P. Murdin (Springer, Berlin, 2017), p. 1
16. T.J. Moriya, E.I. Sorokina, R. Chevalier, Superluminous supernovae. Space Sci. Rev. **214**, 59 (2018)
17. M. Modjaz, Stellar forensics with the supernova-GRB connection. Astronomische Nachrichten **332**(5), 434–447 (2011)
18. M. Fraser, Supernovae and transients with circumstellar interaction. R. Soci. Open Sci. **7**, 200467 (2020)
19. S.E. Woosley, Bright supernovae from Magnetar birth. Astrophys. J. Lett. **719**, L204 (2010)
20. M.P. Allen, J.E. Horvath, Influence of an internal magnetar on supernova remnant expansion. Astrophys. J. **616**, 346 (2004)

21. D. Kasen, L. Bildsten, Supernova light curves powered by young magnetars. Astrophys. J. **717**, 245 (2010)
22. M. Bersten, O.G. Benvenuto, M. Orellana, K. Nomoto, The unusual super-luminous supernovae SN 2011kl and ASASSN-15lh. Astrophys. J. Lett. **817**, L8 (2016). https://doi.org/10.3847/2041-8205/817/1/L8
23. L.I. Sedov, *Similarity and Dimensional Methods in Mechanics* (CRC Press, Boca Raton, 1993)
24. E.A. Dorfi, Evolution of supernova remnants including particle acceleration. Astron. Astrophys. **234**, 419 (1990)

Chapter 6
Astrophysics of Compact Objects

6.1 Formation Events of Compact Objects: Statistics

The theory of Stellar Evolution discussed in Chap. 4 has given us the elements we need to understand the problem that now occupies us: compact stellar remnants. We discussed the evolution of low- and intermediate-mass stars (solar type) and the transition to so-called "high-mass" stars, which proceed to explode after a rapid final evolution. It is important to note that the existence of the two types separated by the mass $8M_\odot$ should also be complemented with an evaluation of the relative number of stars that produce the corresponding compact objects (white dwarfs and neutron stars/black holes). We revisit the whole issue of the theoretical status and observational evidence for compact objects in this Chapter. Figure 6.1 shows the so-called *initial mass function* (IMF), i.e., the number of stars per unit mass (logarithmic) as a function of the mass determined in various studies of the local environment, clusters, and other systems.

The number of progenitors is generally expressed as being proportional to $(M/M_\odot)^{-\alpha}$, and since the pioneering work of E. Salpeter in 1955 [1], the value of the appropriate exponent has been found to be around 2.3. This means that the number of stars that produce white dwarfs is at least 50 times greater than those that explode. Thus, more than 95% of visible stars should form white dwarfs at the end of their evolution. And taking into account a number of complex factors in the evolution of the galaxy, we have come to the conclusion that there may be up to 1 billion white dwarfs available for study.

The statistics of the relative fraction of neutron stars and black holes is much more uncertain. The number of stars that should explode is quite well known, but it is not clear if there is a minimum value from which the production of black holes is inevitable. This stems from the fact that the Physics of explosions for each case does not offer a clear answer. To make matters worse, the initial angular momentum of the collapsing core may turn out to be very important, even fundamental in determining the explosion. There is a certain vague consensus that black holes would form in parent explosions above about $25M_\odot$, by the collapse of the core after ejection when

© The Author(s), under exclusive license to Springer Nature Switzerland AG 2022
J. E. Horvath, *High-Energy Astrophysics*, Undergraduate Lecture Notes in Physics,
https://doi.org/10.1007/978-3-030-92159-0_6

Fig. 6.1 Initial mass function according to several studies as indicated. The scarcity of the number of progenitors that should end their lives with explosions compared to those that will produce white dwarfs (up to roughly $8M_\odot$) is evident [2]. Credit: Johannes Buchner

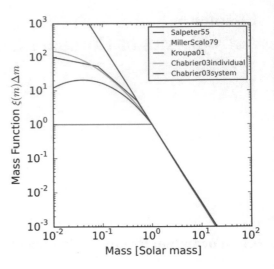

the matter that failed to shut down falls back on top of it (in the process called *fallback*), or directly from imploding stars of around $40M_\odot$ or more [3]. The fact is that in known X-ray binaries (see below and Chap. 7) there is no evidence for black holes of more than around $15M_\odot$, nor for the "very light" ones immediately above the maximum neutron star mass and below around $5M_\odot$, an observation which has been suggested as determined by the very mechanism of the explosions. In the case of explosions forming neutron stars, it is not clear what exactly the formation channels would be. For example, collapse induced by the addition of a white dwarf (AIC) appears as a recurring possibility, but there is no proof of its effectiveness. All this makes it very difficult to evaluate populations, although we usually find 10^7 as an indication of the number of neutron stars in the galaxy (pulsars and others) and something like 10^6 for the black holes produced by stellar evolution [4]. We will have a more accurate picture of this and other issues when we begin to analyze each type of remnant below.

6.2   Theory and Observations of White Dwarfs

6.2.1   In the Beginning . . .

The long history of the study of white dwarfs began with an observation by F. Bessel in 1844. By carefully determining the orbits of Sirius and Procyon, Bessel found that there were systematic periodic deviations, and proposed the existence of undetected "dark companions". In the following decades, some candidates for "dark companions" were finally detected, even at visual magnitudes ≤ 10. In particular, 40 Eridiani B was the object of in-depth study and, to general surprise, Russell, Pickering, and Fleming showed in 1910 that this star was spectral type A, i.e., it had an effective

Fig. 6.2 Contemporary images of the Sirius A and B system in the optical range (*left*) and X-ray (*right*). Credit: NASA, ESA, H. Bond (STScI) and M. Barstow (University of Leicester) (left); NASA/SAO/CXC (right)

temperature in the range 7500–10000 K, considered to be very "white". This did not correspond at all to the expectation for a star of very low brightness [6].

The most obvious conclusion was that these stars were enormously dense, with estimated densities of thousands of times the density of water. Only then could a very low luminosity (remember that $L \propto R^2 T^4$) and a very high temperature be compatible, at the expense of greatly decreasing the radius R. In 1927 A.S. Eddington expressed this strangeness in his characteristically humorous style:

> [...] the message of the companion of Sirius when decoded, runs: "I am composed of matter 3000 times denser than anything you have come across. A ton of my material would be a small nugget you could put in a matchbox". What reply can one make to such a message? The reply which most of us made in 1914 was: "Shut up. Do not talk nonsense".

Eddington implicitly recognizes in this paragraph the need to apply new ideas to study the behavior of matter at these densities. Evidently, the classical gas approach cannot work in this situation, and it was the work of R.H. Fowler in 1926 which laid the foundations for the treatment of the problem of the structure of Sirius B (Fig. 6.2) and other white dwarfs, a name suggested by the necessary temperature and radius. It is important to point out that modern Quantum Mechanics had been completely formulated only two years before in 1924. Thus we have a perspective on the revolutionary nature of these initial studies of white dwarfs, which are "natural" laboratories of dense matter and led to one of the great physical achievements of the new quantum approach.

6.2.2 Matter in the High Density Regime ($\rho \geq 10^3 \, \text{g cm}^{-3}$)

As we said before, the inference of very high densities, where an ideal gas model would not be viable, forced astrophysicists to consider the behavior of matter in the regime already presented in Figs. 4.17 and 4.18. We will now see how it is possible to obtain and justify a valid *equation of state* for that regime from elementary considerations.

Let us consider once again the situation of having N electrons confined in a volume V. The physical space available for each of them is in one dimension of the order of $\Delta x \sim (V/2N)^{1/3}$. The hypothesis that electrons are in the quantum regime is equivalent to saying that they are now subject to the Uncertainty Principle $\Delta x \Delta p \geq \hbar$, so their typical momentum will be of the order of

$$\Delta p \geq \frac{\hbar}{2\Delta x} \approx \frac{\hbar N^{1/3}}{2^{2/3} V^{1/3}} \, . \tag{6.1}$$

The mean kinetic energy $\langle E_K \rangle$ would be

$$\langle E_K \rangle = \frac{\Delta p^2}{2m} \approx \frac{\hbar^2 N^{2/3}}{2^{7/3} V^{2/3} m} \, . \tag{6.2}$$

Therefore, the internal energy U is simply

$$U = N \langle E_K \rangle \approx \frac{\hbar^2 N^{5/3}}{2^{7/3} V^{2/3} m} \, . \tag{6.3}$$

This last relation is important for the following reason: in a totally general way, Thermodynamics allows us to find the pressure (gas state variable) by differentiating the internal energy with respect to the volume at constant entropy, since the internal energy is one of the Thermodynamics potentials of the system. Hence, $P = -\partial U/\partial V|_{S=\text{const.}}$. Thus, we have

$$P = \frac{\hbar^2 N^{5/3}}{2^{4/3} 3 V^{5/3} m} \, , \tag{6.4}$$

i.e., $P \propto n^{5/3}$ since the particle number density is $n = N/V$, as we anticipated in Chap. 4. This also reveals an important fact: as the Planck constant squared $\hbar^2$ appears in the expression, this pressure due to degeneracy would *not exist* without Quantum Mechanics. All these facts are present in Fowler's 1926 paper [7].

Note that the calculation is simple and completely general. If we had considered ultra-relativistic electrons with $\langle E_K \rangle = pc$, we would have obtained $P \propto n^{4/3}$ instead. These two forms are the high and low density limits of the degenerate electron gas and will be useful in the calculation of white dwarf structure, as we will see below.

6.2.3 White Dwarf Structure

As discussed in Chap. 4, and assuming that the degenerate matter that constitutes the white dwarf does not produce energy by means of nuclear reactions, the structure of these stars can be found by simultaneously integrating the equations of continuity of mass and hydrostatic balance. The energy transport equation is also dispensable, since degenerate electrons have a very high conductivity and thus $dT/dr = 0$. It is assumed that the interior temperature is constant for this reason, except for the outer layers where degeneracy ends and the gas is "normal" again, and the interior temperature drops until it reaches the value of the surface, where there is black body emission.

As in any system of two first order differential equations, we can combine dM/dr and dP/dr to obtain an equivalent second order equation. This unique differential equation is

$$\frac{1}{r^2}\frac{d}{dr}\left(\frac{r^2 dP}{\rho dr}\right) = -4\pi G\rho .$$ (6.5)

We see that, as in more general cases, we require a relationship between P and ρ (the equation of state, see Chap. 4), just like the one obtained for a degenerate electron gas. For the purpose of a general treatment, a *polytropic* form $P = K\rho^{\Gamma}$ is usually introduced, a general case comprising the limits $P \propto n^{5/3}$ and $P \propto n^{4/3}$ relevant to our case. Certain mathematical manipulations can render the problem more tractable. For example, the exponent Γ of the polytropic equation of state can be replaced by another by writing $\Gamma = 1 + 1/n$, where n is called the *polytropic index*.

We define a change of variables in (6.5) by

$$\rho = \rho_c\Theta^n , \quad r = a\xi , \quad a = \left[\frac{(n+1)K\rho_c^{1/(1-n)}}{4\pi G}\right]^{1/2} .$$ (6.6)

With this change we can now get (6.5) in a dimensionless form, viz.,

$$\frac{1}{\xi^2}\frac{d}{d\xi}\left(\frac{\xi^2 d\Theta}{\rho d\xi}\right) = -\Theta^n .$$ (6.7)

This is known as the *Lane–Emden equation* in honor of the scientists who studied it. Besides the formal problem of finding solutions, we should not forget that we are looking for a description of white dwarfs. Thus, the boundary conditions imposed by the Physics of the problem for the solution function $\Theta(\xi)$ are quite simple:

$$\Theta(\xi = 0) = 1 , \quad \Theta'(\xi = 0) = 0 .$$ (6.8)

The first comes from the fact that $\rho(r = 0) = \rho_c$, and the second describes the fact that $dP/dr = 0$ at the center, otherwise we would have pressure gradients (forces) where $M \approx 0$.

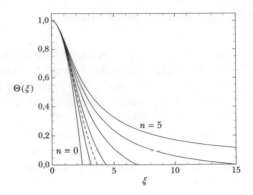

Fig. 6.3 Exact solutions of the Lane–Emden equation (1.9) for polytropic state equations parametrized by the polytropic index $n = 0$, 1, 3/2, 3, 5

In general, the solutions of (6.7) decrease from a central value to a point ξ_1 that crosses the horizontal axis where $\Theta(\xi_1) = 0$ (Fig. 6.3). This point is of interest to us because we identify it with the star radius R, since it is where $P = 0$. The stellar radius R can be expressed in general in terms of ξ_1 by

$$R = a\xi_1 = \left[\frac{(n+1)K}{4\pi G}\right]^{1/2} \rho_c^{(1-n)/2n}\xi_1 , \qquad (6.9)$$

and the mass contained inside this radius is

$$M = 4\pi a^3 \xi_1^2 \rho_c |\Theta'(\xi_1)| . \qquad (6.10)$$

Explicit construction of the solutions of the Lane–Emden equation is only possible for certain particular values of the index n, e.g., $n = -1, 0, 1, 3/2, 3, 5$, and ∞, and yet only for $n = 0$, 1, and 5 can the solution functions be constructed analytically. In some cases the condition for existence of zero at $\xi = \xi_1$ is not satisfied and it is not possible to build viable star models, since these lack "radius" if Θ does not cross the horizontal axis. In general, the profiles of Θ are very concentrated in the center, i.e., they differ a lot from approximations with $\rho = $ constant. A detailed and explicit treatment of all these cases and other related issues can be found in the classic reference by Chandrasekhar [8], while the relativistic generalization can be found in [9].

In possession of some general solutions of the Lane–Emden equation (Fig. 6.4), and assuming for the time being that the functions Θ are well behaved and have a zero at $\xi = \xi_1$ for any useful value of the index n, we can, by manipulating the formulas (6.9) and (6.10), find the general relationship between mass and radius for the polytropic model:

$$M = 4\pi R^{(3-n)/(1-n)} \left[\frac{(n+1)K}{4\pi G}\right]^{n/(n-1)} \xi_1 |\Theta'(\xi_1)| . \qquad (6.11)$$

Therefore, the mathematical solutions allow us to build a whole sequence of white dwarf models. In the limiting cases of interest (non-relativistic and ultra-relativistic electrons), we have

$$\Gamma = 5/3 \ (n = 3/2): \quad \xi_1 = 3.65375, \quad \xi_1^2 |\Theta'(\xi_1)| = 2.71406, \quad (6.12)$$

$$\Gamma = 4/3 \ (n = 3): \quad \xi_1 = 6.89685, \quad \xi_1^2 |\Theta'(\xi_1)| = 2.01824. \quad (6.13)$$

When physical units are restored, the results for low-density white dwarfs with $\Gamma = 5/3$ ($n = 3/2$) are

$$R = 1.22 \times 10^4 \left(\frac{\rho_c}{10^6 \, \text{g cm}^{-3}} \right)^{-1/6} \left(\frac{\mu_e}{2} \right)^{-5/6} \text{km}, \quad (6.14)$$

$$M = 0.4964 \left(\frac{\rho_c}{10^6 \, \text{g cm}^{-3}} \right)^{1/2} \left(\frac{\mu_e}{2} \right)^{-5/2} M_\odot = 0.7 \left(\frac{R}{10^4 \, \text{km}} \right)^{-3} \left(\frac{\mu_e}{2} \right)^{-5} M_\odot, \quad (6.15)$$

while for the high-density case with $\Gamma = 4/3$ ($n = 3$), we have

$$R = 3.347 \times 10^4 \left(\frac{\rho_c}{10^6 \, \text{g cm}^{-3}} \right)^{-1/3} \left(\frac{\mu_e}{2} \right)^{-2/3} \text{km}, \quad (6.16)$$

$$M = 1.457 \left(\frac{\mu_e}{2} \right)^{-2} M_\odot \quad \text{(independent of } R!\text{)}. \quad (6.17)$$

Figure 6.4 compares Chandrasekhar's polytropic models, whose limits correspond to the equations of state of this section.

We now have analytical models for the white dwarfs that tell us that the radii *increase* as the mass is reduced (!), a fact that results from the degenerate nature of the matter and is quantified in the relation (6.15). Moreover, in the high density limit we reach a maximum of the mass for the stars of this sequence, and for this maximum mass the minimum of the radius (since the maximum mass corresponds to the maximum density). The most massive white dwarfs have radii comparable to the Earth, and "pack in" almost one and a half solar masses. A graphical comparison is shown in Fig. 6.5. It is not strange that the photon emission (proportional to R^2) is weak, as observed.

As a final comment, (6.16) and (6.17) imply that, when $\rho_c \to \infty$, the mass of the star tends to the maximum value $1.457(2/\mu_e)^2 M_\odot$, known as the *Chandrasekhar limit* [4]. We interpret this as the maximum mass that a star can have supported by the pressure of degenerate electrons. This result of the existence of a maximum is quite surprising and was not accepted without difficulty by researchers at the beginning of the 20th century (in particular by A.S. Eddington, although he later acknowledged the limit). However, it is today very well established and constitutes one of the pillars of contemporary Astrophysics. Therefore, it will be important to understand its broad meaning, as we shall attempt to now.

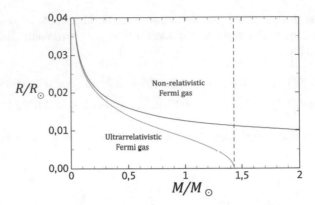

Fig. 6.4 Stellar sequences obtained by the integration of the Lane–Emden equation. In *blue*, the sequence of star models built with the non-relativistic limit of the Fermi gas of electrons that satisfies $P \propto n^{5/3}$. As expected, deviations are increasingly important as mass increases. At some intermediate point ($M \sim 0.6$–$0.7 M_\odot$), one must shift to the description with the ultra-relativistic limit, where $P \propto n^{4/3}$ (*green curve*), whose results are more and more accurate until it reaches the value where its derivative becomes vertical (*red dotted line*). There are no stable models beyond this value, the Chandrasekhar limit

Fig. 6.5 Graphic illustration of the relative size of a high-density white dwarf and the Sun, in approximately the right scale

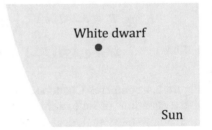

6.2.4 *Chandrasekhar Limit*

For its multiple applications, the so-called *Chandrasekhar limit* (again, not to be confused with the Schoenberg–Chandrasekhar limit presented in Chap. 4, which refers to another physical situation) is one of the most important results obtained in the 20th century for the theory of Stellar Evolution. It is also highly significant that it depends in a fundamental way on the ideas of Quantum Mechanics, very new at the time of Chandrasekhar's original work. We have already seen in Chap. 4 that the concept of degeneracy is fundamental for the evolution of solar-type star cores. Without this state, there would be no helium flash, for example. The theoretical physicist Lev Landau reasoned that, since Chandrasekhar's mass limit is so fundamental, he should be able to demonstrate it with very simple arguments (he would be amazed by the 3D numerical simulations, etc., that are made today precisely with the intention of discovering fundamental results). We will now present Landau's argument, because it will allow us to understand Chandrasekhar's mass limit in a qualitative and simple way [4].

Consider once again N fermions contained in a sphere of radius R. The number density of the fermions is $n \sim N/R^3$, so by the Heisenberg principle the momentum of each fermion should be $p \sim \hbar/V^{1/3} = \hbar n^{1/3}$, since they are confined to this volume V. The corresponding Fermi energy is $E_F \sim pc \approx \hbar c N^{1/3}/R$ in the non-relativistic limit. Each fermion also feels a gravitational energy $E_g \approx GMm_B/R \approx GNm_B^2/R$, product of the gravitational attraction that exerts all the mass distribution on it.

As in any physical system, the equilibrium state of the configuration is reached at a minimum of the total energy $E = E_F + E_g$, i.e., we should minimize

$$E = \frac{\hbar c N^{1/3}}{R} - \frac{Gm_B^2 N}{R}. \tag{6.18}$$

However, there is a important feature regarding the existence of this minimum related to the number of N fermions. If N is small, the first term dominates (since $N^{1/3} > N$) and E is positive. Thus, one can decrease the energy by increasing R. When the star expands to decrease the energy, the fermions will at some point become non-relativistic particles ($E_F \to p_F^2/2m \propto 1/R^2$) and the second term will now dominate, making $E \to 0^-$, i.e., it will tend to zero from negative values, before if $R \to \infty$. Therefore, there must be a point of equilibrium for a *finite value* of the radius R (that is, a star). But if we consider N large enough from scratch, E will be negative and tend to $-\infty$ if $R \to 0$, that is, the configuration will *collapse*, because it will then be able to decrease the energy indefinitely, and there is no possible equilibrium. The boundary between "small" N and the "large" N that separates these two regimes corresponds to a maximum value of fermions $N_{\max}$, precisely determined by the condition $E = 0$ in (6.18), and which is easily calculated to be

$$N_{\max} \approx \left(\frac{\hbar}{Gm_B^2}\right)^{3/2} \sim 2 \times 10^{57}. \tag{6.19}$$

The associated maximum mass $M_{\max} \equiv M$ when $N = N_{\max}$ is then

$$M_{\max} \approx N_{\max} \times m_B \sim 1.5 M_\odot. \tag{6.20}$$

Note that $N_{\max}$ and $M_{\max}$ depend essentially on *universal constants*, not on the composition, since that was never needed in the argument. We can also show that the *equilibrium radius* is determined by the condition of the onset of relativistic degeneracy, viz.,

$$E_F \approx mc^2, \tag{6.21}$$

where m is the mass of the particle whose pressure supports the star. Substituting for $N_{\max}$ in $E_F \approx \hbar c N_{\max}^{1/3}/R_{\max}$ and using the condition (6.19), we obtain

$$R_{\max} \approx \frac{\hbar}{mc} \left(\frac{\hbar}{Gm_{\mathrm{B}}^2} \right)^{1/2} . \tag{6.22}$$

This result suggests two distinct regimes:

- if $m = m_{\mathrm{e}}$ (white dwarfs), $R_{\max} \approx 5 \times 10^8$ cm,
- if $m = m_{\mathrm{neutron}} = m_{\mathrm{B}}$, i.e., the neutrons themselves become degenerate, the value will be 2000 times smaller, viz., $R_{\max} \approx 3 \times 10^5$ cm (neutron stars).

Therefore we have *two* regimes of stable degenerate objects: the less dense of them reaches the condition of instability for

$$\rho_c \approx 3M/4\pi R^3 = \frac{4.5(2 \times 10^{33})\,\mathrm{g}}{4\pi \times 125 \times 10^{24}\,\mathrm{cm}^3} \sim 10^8\,\mathrm{g/cm}^3$$

giving an order of magnitude for the maximum density for white dwarfs, while the densest becomes unstable above $\rho_c \approx 10^{15}$ g/cm³, given an order of magnitude for the maximum density for neutron stars. However, we must remember that in the latter the effects of General Relativity and the interactions between particles are important, and our basic Newtonian estimate presented here will not be very reliable. In fact, it should be noted that we obtained as a result the existence of a maximum mass *without using GR concepts*, while at high densities we will have to deal with relativistic instability, the true cause of the maximum mass of a compact object in the neutron matter regime. On the other hand, we can say that the maximum mass should be approximately the same for both regimes of (6.18), within a small numerical factor that cannot be determined by this simple calculation.

6.2.5   Observations of White Dwarfs

The fact that the white dwarfs were detected at the beginning of the 20th century is another proof of the statement made at the beginning of the Chapter regarding their abundance in the galaxy. Sirius B, 40 Eridiani B, and other binary white dwarfs are examples of the presence of these objects in the vicinity of the Earth. There are many others, most of them isolated, and some with magnitudes less than 12, accessible to any amateur telescope. It is not difficult to find and observe white dwarfs.

But of course the systematic study of white dwarfs requires large samples and as complete as possible. Thus, in addition to neighboring and "field" white dwarfs, there are studies of old star populations, each of approximately the same age, responsible for the production of white dwarfs: the star clusters, which are especially suitable laboratories [10].

Figure 6.6 shows the case of the NGC 6791 cluster, where colors and luminosities are used to identify the white dwarfs born from solar type progenitors that have already completed their evolution. With samples of this type it is possible to study

Fig. 6.6 White dwarfs in the globular cluster NGC 6791. With high quality images the identification is quite simple (the candidates are the *dots in the circles*), and one can extend the study by obtaining complementary spectra. Credit: NASA, ESA, and L. Bedin (STScL)

the white dwarfs and related problems, such as the determination of the very age of the cluster through the sample of white dwarfs.

All these properties like colors and spectra still require a detailed treatment of the atmospheres of the white dwarfs, a region totally ignored in our discussion of the structure because it is represents such an insignificant fraction of the total mass. After all, it is responsible for the radiation that is ultimately emitted by these objects. Figure 6.7 shows the situation graphically.

Observation of white dwarfs has the potential to determine several important characteristics of their structure, for example, the theoretically calculated star radius. If we call the observed luminous flux $F(D)$, the basic equation $L = 4\pi R^2 \sigma T_{\text{eff}}^4$ can in principle be used to obtain the stellar radius:

$$F(D) = \frac{L}{4\pi D^2} \implies R^2 = \frac{FD^2}{\sigma T_{\text{eff}}^4} \, . \tag{6.23}$$

We see from (6.23) that, besides the distance D, we must determine the effective temperature T_{eff}. Although this is not impossible, there are several complications, as exemplified in Fig. 6.8. The spectra of many white dwarfs have pronounced absorp-

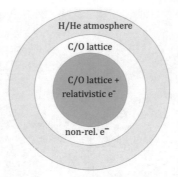

Fig. 6.7 Complete structure of a white dwarf. The polytropic treatment presented above is valid for most of the matter in the white dwarf, but not for the atmosphere, which hardly contributes to the mass but which is where the degeneracy of the electrons ends and there is a transition to a classical gas. Besides confirming the most common composition for the atmospheres (H/He), we will see that more relevant situations are observed

Fig. 6.8 Spectrum of a white dwarf showing multiple identified lines (this absorption is called *blanketing* in astronomical jargon), something which introduces an uncertainty in the determination of the effective temperature T_{eff} [11]. Credit: Detlev Koester

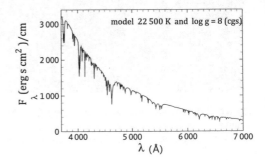

tion lines that distort the spectrum with respect to the ideal, and thus make it difficult to calculate the value of T_{eff}. Here we see another advantage of the study of clusters: the distance D is the same for all objects.

With the construction of increasingly complete databases, it has been possible to classify white dwarfs using their spectra. This classification is shown in Table 6.1. There are complicated evolutionary mechanisms that result in the transformation of some types into others, but they will not be discussed here, since they involve the Physics of the diffusion of chemical elements and other problems that go well beyond the scope of this text.

However, we would like to highlight some novelties and an important recent contribution to this problem: the DQ class with carbon lines was only recently discovered, although its existence was expected. But finding a white dwarf with oxygen and without hydrogen or helium was not expected at all. This is the case of SDSS J124043.01+671034.68, discovered by Kepler, Koester, and Ourique [11], and which should result from the more massive progenitors that still cannot explode. In fact, the detection of neon and magnesium in the oxygen-rich atmosphere points to the white dwarf coming from this type of core, very close to those that will produce electron-

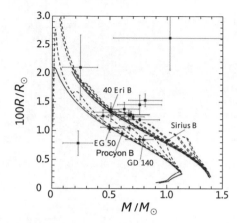

Fig. 6.9 Theoretical mass–radius diagram for nearby white dwarfs. The *upper curve* corresponds to a carbon composition, and the lower one to iron. According to the theory of Stellar Evolution, it is impossible to have iron white dwarfs, although this would appear empirically to be the indicated solution. For this reason, a re-evaluation of distances is indispensable, since the observed points are expected to migrate vertically and correspond to theoretical expectations [12]. © AAS. Reproduced with permission

Table 6.1 Empirical spectral classification of white dwarfs. The scheme is conceptually similar to the spectral type classification created for "normal" stars, and reflects the previous evolutionary history of each object

Spectral type	Features
DA	Only H, no He I or metals
DB	Only He I, no H or metals
DC	Continuous spectrum, no lines
DO	Strong He II, He I or H may appear
DZ	Only metals, no H or He I
DQ	C lines of any type

capture supernovas (Chap. 5). There is still no spectral denomination for this unusual object. The location of several white dwarfs in the M-R plane is displayed in Fig. 6.9.

One result of the above-mentioned studies is confirmation of the mass of progenitors that produce white dwarfs, as in the discussion above. NGC 2751 is an open cluster with a white dwarf as a member. This membership is quite reasonable for a white dwarf with a hydrogen atmosphere. The important fact is that the cluster has quite massive stars that are still in the Main Sequence. Thus, the progenitor of the white dwarf should be more massive than those that have still to begin their final evolution (Fig. 6.10).

This work indicates that the basic ideas of Stellar Evolution are not terribly wrong. Other cases of (multiple) white dwarfs in clusters have been published, and imply somewhat lower limits to the progenitor mass in the Main Sequence, probably as

Fig. 6.10 Main Sequence in the cluster NGC 2751 [13]. Superposition on the theoretical sequences shows that stars of mass around $7.5M_\odot$ are still in it. Thus, we know that the progenitor of the white dwarf had mass greater than $7.5M_\odot$. Credit: E.E. Giorgi et al., Astron. Astrophys. **381**, 884 (2002), reproduced with permission © ESO

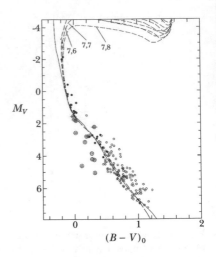

the result of a different metallicity (but certainly all above $6M_\odot$). A great source of uncertainty in this problem is the *mass loss* in the giant branch and/or the AGB, a factor that could cause stars that should explode to form white dwarfs (maybe even around $10M_\odot$). On the other hand, if the largest mass that forms white dwarfs were too low, there would be a very serious conflict with the number of supernovas observed.

Another important issue in the study of white dwarfs is their mass distribution. Internal composition and mass are expected to increase as we consider white dwarfs that descend from more massive progenitors. However, in the low-mass range, it is not possible to produce helium white dwarfs, since the cores merge this helium into carbon (Chap. 4). There is a consensus in favor of the production of helium white dwarfs but in *binary systems* only. And, as we have already said, those of greater mass should be composed of oxygen with fractions of neon and magnesium, and masses close to the Chandrasekhar mass.

Figure 6.11 shows the mass distribution obtained by Kepler et al. [14]. The maximum around $0.57M_\odot$ is very similar to the one obtained in other studies. There are secondary maxima, tentatively associated with several formation channels, such as He-light white dwarfs formed in binary systems, as already mentioned. At the right-hand end of the histogram, one can see a white dwarf with mass $1.33M_\odot$, quite close to the Chandrasekhar limit. There are other cases of even higher mass, but they are subject to confirmation.

6.2.6   Cooling and Crystallization of White Dwarfs

The absence of nuclear reactions in the white dwarfs indicates that, from the moment they come into existence, they can only shed their thermal energy content. Therefore, a cooling theory must be formulated to study the population of the Galaxy as a whole.

Fig. 6.11 Mass distribution obtained by Kepler et al. Several maxima appear, possibly associated with different formation channels. Credit: Fig. 8 of [14]

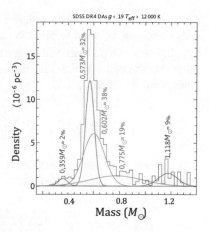

Fortunately, thanks to their structural simplicity, the required elementary cooling theory is very simple. We start with the definition of stellar luminosity, identified with the variation of the internal thermal energy E_{th}, and use the chain rule to bring in the temperature T_c of the isothermal core:

$$L = -\left(\frac{\partial E_{th}}{\partial T_c}\right)\left(\frac{\partial T_c}{\partial t}\right) . \tag{6.24}$$

The first term in brackets is actually the *specific heat* of the reservoir. Although we have seen that it is the electrons that maintain the structure, their contribution to the thermal energy is very small. The thermal reservoir is largely dominated by the classical ions, so we write $c_V^{ion} = (3/2)N_A k_B/A$. Hence, we have

$$L = -6.4 \times 10^7 \frac{1}{A}\frac{M}{M_\odot}\left(\frac{\partial T_c}{\partial t}\right) . \tag{6.25}$$

Now the luminosity is a function of the central temperature variation T_c, which needs to be evaluated. For this we will consider the white dwarf envelope, which contains a very small mass but is the region where the temperature falls from the inner value T_c to the final value T_{eff} at the photosphere. With the hypothesis that the envelope mass is $M_{env} \approx 0$, that is, that it does not contribute to the total mass, we can divide the transport equation by the hydrostatic balance equation to obtain

$$\frac{dT_c}{dP} = \frac{3}{4ac}\frac{L}{4\pi GM}\frac{\bar{\kappa}}{T^3} . \tag{6.26}$$

In the envelope, in the interphase region we mentioned, the matter ceases to be degenerate and its opacity is dominated by processes that have a Kramers form (Chap. 4)

$$\bar{\kappa} = \kappa_0 \rho T^{-7/2} . \tag{6.27}$$

Here, we assume that the degenerate pressure and the normal gas pressure are the same, because we want to find the conditions for the transition. This yields a relationship between pressure and temperature, which results in $P_c \approx T_c^{5/2}$. Inserting the latter and the opacity (6.27) in (6.26), we can separate variables and integrate both sides, with the result

$$\frac{L}{L_\odot} = 1.7 \times 10^{-3} \frac{M}{M_\odot} \frac{\mu}{\mu_e^2} \frac{4 \times 10^{33}}{\kappa_0} \left(\frac{T_c}{10^7 K} \right)^{7/2} . \tag{6.28}$$

The last step is to substitute (6.28) into (6.25) and integrate over time to obtain the time required to achieve a given luminosity:

$$t_{\text{cool}} = 9 \times 10^6 \mu^{-2/7} \frac{A}{12} \left(\frac{M}{M_\odot} \right)^{5/7} \left(\frac{\mu_e}{2} \right)^{4/3} \left(\frac{L}{L_\odot} \right)^{-5/7} \text{yr} . \tag{6.29}$$

This result is due to L. Mestel [15] and constitutes the simplest cooling theory. We can observe two very interesting characteristics of the expression obtained. The first is that the cooling time t_{cool} is inversely proportional to the atomic number of the ions, i.e., the lighter composition white dwarfs cool more slowly for a given mass. But if we consider white dwarfs of higher mass, they will cool down more slowly since they have higher thermal content and are more compact for a given temperature, whence their emission surface is smaller. Note that effects that may become important, such as the emission of neutrinos from the interior in addition to the luminosity of photons from the surface, have not been included. There is an uncertainty of about 20% due to these factors and other simplifications used to obtain Mestel's law (6.29).

There are several possible tests of cooling. One of the more interesting ones is to find the cooling sequence in a cluster, since the stars are believed to have essentially the *same lifetime*. Figure 6.12 shows the data for the M4 cluster. The white dwarfs

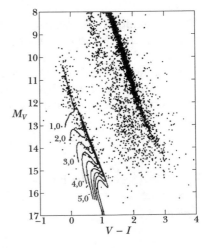

Fig. 6.12 The sequence of white dwarfs in M4 [16]. An absolute minimum magnitude of $M_V \sim 16.5$ suggests a limit for the age of the disk, set by the absolute limit for the age of the oldest white dwarf in the sample. © AAS. Reproduced with permission

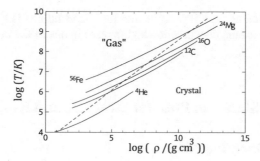

Fig. 6.13 Regions in the $T-\rho$ plane where the state of matter changes as cooling progresses. Since its birth, the white dwarf core has been at high temperatures, and this results in a "gas" (with corrections to the ideal expressions, but still fluid). Below the *dashed line*, the parameter Γ exceeds the critical value, and carbon, oxygen, or even magnesium crystallize. It is still unclear whether the geometry of the crystal is analogous to that of the terrestrial diamond (a cubic net) or something much more exotic (a triangular net, never seen in laboratories)

are clearly separated, below the Main Sequence. The faintest magnitudes observed correspond to luminosities of order $10^{-4}L_\odot$. An important point is that there could be no fainter white dwarfs, since the galactic disk is not old enough for this to happen. Thus, a number of papers have suggested calculating the age of the galactic disk using precisely the cooling of white dwarfs. The results are varied, but oscillate around 6–8 Gyr, which is consistent with other, independent arguments.

To conclude, we highlight an aspect of cooling that has not entered the discussion above, but for which there is substantial evidence: it is the so-called *crystallization* of the core material, expected at low temperatures. This crystallization is due to the fact that, at high temperatures, thermal agitation keeps the ions in a fluid state (referred to as a "gas" in Fig. 6.13). But if the temperature is low, Coulomb interactions of the charged ions can locate the ions in the sites of a crystal lattice. A quantitative criterion, obtained by studying numerical simulations of crystallization, is that the quotient of two quantities reaches a value of about 180, i.e.,

$$\Gamma = \frac{(Ze)^2/\langle r \rangle}{k_B T_c} = 2.3 \frac{Z^2}{A^{1/3}} \frac{\left(\dfrac{\rho}{10^6 \, \text{g cm}^{-3}}\right)^{1/3}}{\dfrac{T_c}{10^7 \, \text{K}}} \approx 180 \,. \tag{6.30}$$

When this condition is reached, the cooling regime changes, since crystallization releases latent heat [4]. This latent heat makes the time t_{cool} increase, since it contributes to E_{th} in (6.24). Thus, white dwarfs in the process of crystallization (from the inside out) almost cease to cool down.

It is worth asking what evidence there is for such crystallization. Studies of the oscillations of white dwarfs allow us indirectly to explore their interior (as is done with terrestrial seismology). In particular, the oscillations of the white dwarf BPM 37093 of only 4 millimagnitudes (!) have been used, after a fit to theoretical calculations, to argue that at least 50% of its interior is crystallized [17]. There are other examples of this extreme phenomenon taking place in one of the most hidden places in the Universe.

6.3 Neutron Stars and Pulsars: Structure and Evolution

6.3.1 The Pioneering Ideas

Neutron stars were "born" in science with an intuition of L. Landau in 1931, in a study he did not publish until after the discovery of neutron, but which in fact preceded it. In "On the theory of stars" [18], Landau speculates about the possibility that gravitation compresses matter and thus forms stars that look like a kind of giant atomic nucleus (see [19] for a complete discussion of this contribution). In this work, published in 1932, months after Chadwick announced the discovery of the neutron, Landau already presents an outline of the structure and estimates the maximum mass of these compact objects. Another very inspired contribution in this respect is due to W. Baade and F. Zwicky who, in 1934, associated supernovas with the birthplace of neutron stars for the first time, indicating a specific place for the compression that Landau needed. Baade and Zwicky wrote [20]:

> With all reserve we advance the view that a super-nova represents the transition of an ordinary star into a neutron star, consisting mainly of neutrons. Such a star may possess a very small radius and an extremely high density.

This statement was published shortly after Landau refined his view of this subject, showing that the densities involved must exceed those of the atomic nucleus and that the maximum mass would be limited, in the case of a degenerate free neutron gas, by the value $0.7M_\odot$, a factor of about 2 lower than originally predicted. Like white dwarfs, neutron stars provide an extraordinary example of the role played by Quantum Mechanics in contemporary Astrophysics, now in the high-density regime (and just before producing black holes). These calculations were repeated later with the help of the relativistic structure equations of Tolman, Oppenheimer, and Volkoff (or TOV, obtained in 1939) and were confirmed to a large extent. Following the example of the hydrostatic equilibrium equation (4.4), these researchers [21] obtained a version that includes the effects of General Relativity:

$$
\frac{\mathrm{d}P}{\mathrm{d}r} = -\frac{G\left(\rho + \dfrac{P}{c^2}\right)\left(M + 4\pi r^3 \dfrac{P}{c^2}\right)}{r^2\left(1 - \dfrac{2GM}{rc^2}\right)} .
\tag{6.31}
$$

This reduces to the Newtonian version provided that the terms of the P/c^2 (which do not contribute to the gravitational field in Newtonian gravity, but are present in GR) are discarded and we adopt the non-relativistic approximation $2GM/rc^2 \ll 1$. This is precisely where we see the reason why it is necessary to use GR: while for an ordinary star, or even for a white dwarf, the dimensionless quantity $2GM/rc^2 \sim 10^{-4}$, one generally has $2GM/rc^2 \approx 0.1$ for a neutron star, and this factor cannot be neglected. The conservation equation $dM/dr = 4\pi\rho r^2$ has the same form in GR, although its meaning is different [22]. Of course, (6.31) is much harder to solve than (4.4), since these terms make the mathematics very difficult, and we also need an equation of state $P(\rho)$. However, there are some simple models that respect all that is "desirable" for a physically relevant solution (finite at the origin, existence of a zero for the radial coordinate, etc.). These were studied by R. Tolman himself [23], as well as many other researchers over almost a century [24].

6.3.2 Matter in the Neutronization Regime ($\rho \geq 10^{11}\,g\,cm^{-3}$)

The question of the equation of state can be approached in a first approximation in the same way as Landau did, using the expression for a degenerate gas (1.4). In the high-density regime, degeneracy corresponds to neutrons themselves, since almost all electrons will be captured by protons and matter will be strongly *neutronized*. However, the factor $1/m$ indicates that when we substitute the mass of the electron for the mass of the neutron the pressure should drop substantially, and with it the maximum mass. However, in the high density regime, with the structure described by (6.31), there is another feature which becomes much more important than degeneracy: the presence of the pressure in the TOV equation produces a *relativistic instability* when the mass of the star grows. Thus, there is also a *limiting mass*, but which is *not at all* related to the Chandrasekhar mass. This maximum mass may be called the *TOV mass*. One can see that there is a very big difference between these two concepts: for densities above nuclear saturation $\rho_0 = 2.7 \times 10^{14}$ g cm^{-3}, the neutrons are so close together that nuclear interactions cannot be ignored. Thus, the neutron matter in most of the star is much "harder" than the neutron-free gas, i.e., it exerts more pressure for the same energy density, so it is these repulsive interactions that determine the value of the TOV mass. This value must be at least $2M_\odot$ to be compatible with some observed masses, as we will see below. Thus, the question of the equation of state becomes the main one, and we need to discuss the state of matter in a realistic way, going beyond the simple neutron gas.

The description of matter become increasingly difficult beyond the density of the white dwarf centers, as uncertainties increase in the ultra-dense regime. As we have already pointed out in Chap. 5, if the density exceeds the value of $\rho_{\mathrm{drip}} \approx 4 \times 10^{11}$ g cm^{-3}, it is favorable for neutrons to leak out from the nuclei, and not remain as a component of them. This *neutron drip point* marks the beginning of the presence of a free neutron gas and the equation of state then begins to feel the contribution of free neutrons, which already make a substantial difference to the pressure at $\rho > 10^{12}$ g cm^{-3}, and dominate beyond about 10^{13} g cm^{-3}. The equation of state in this regime is typically modeled by starting with a semi-empirical mass function for the nuclei, based on the liquid drop model, of the type

$$Mc^2 \equiv E = -\epsilon_0 A + \epsilon_S A^{2/3} + \epsilon_C Z^2 A^{-1/3} .$$

In this approach the nucleus is treated as a "little drop" of matter and its energy (mass times c^2) is assumed to be composed of a volume term (the first, proportional to the number of nucleons A), another associated with the surface (the second, proportional to $A^{2/3}$), Coulomb corrections (third term), and other effects of minor importance not shown here. The task here is to adjust the expression to reproduce masses of known nuclei and then obtain the coefficients ϵ_0, ϵ_S, ϵ_C. The next step is to calculate what happens at the high densities including these nuclei, the neutron gas, etc. The equation of state (EoS) is obtained by first minimizing with respect to A and Z, and imposing chemical and mechanical equilibrium between the neutron gas and the nuclei. These four conditions allow us to express the total energy density ϵ as a function of a single variable (usually the baryon density n_B), and then obtain the pressure in the form

$$P = n_B^2 \frac{\partial}{\partial n_B} \left(\frac{\epsilon}{n_B} \right) = P_n + P_e + P_L . \tag{6.32}$$

The prototype of this calculation was carried out by Baym–Bethe–Pethick (BBP) [25] to describe matter in this range of densities, valid approximately up to the saturation density [4]. At the saturation point, the nuclear structure dissolves and the matter is composed to a first approximation of fluids of neutrons, protons, and electrons. There the n–p, p–p, and n–n interactions become important. A possible (non-relativistic) approach that aims to treat nucleon–nucleon forces in the simplest possible way consists in writing a generalization of the Yukawa potential:

$$V_{BJ} = \Sigma_j C_j \frac{e^{-j\mu r}}{\mu r} + V_T , \tag{6.33}$$

where μ is related to the reciprocal of the mass of the particle (generically called a *meson*) that is exchanged between the nucleons (Chap. 1), and V_T are additional (tensor) terms in the potential. In their classic work, Bethe and Johnson [26] besides the attractive interactions due to the exchange of pions, considered the dominant effect of the vector meson ω, largely responsible for the repulsive core, to contribute a term $V_\omega = g_\omega^2 e^{-\mu_\omega r}/r$, where $g_\omega^2/\hbar c \sim 29$, as derived from laboratory scattering data. The Bethe–Johnson EoS, which they called model I, is obtained by combining

$$\frac{\epsilon}{n_B} = 236 n_B^{1.54} \text{ MeV/particle} + m_p c^2 , \tag{6.34}$$

$$P = n_B^2 \frac{\partial}{\partial n_B} \left(\frac{\epsilon}{n_B} \right) = 364 n_B^{2.54} \text{ MeV/fm}^3 . \tag{6.35}$$

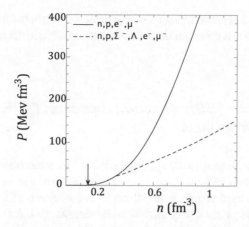

Fig. 6.14 Example of the difference between state equations calculated in the ultra-dense regime. The *upper curve* is derived by considering only n, p, e⁻, and μ⁻. The *bottom curve* includes Λ, which behaves as a kind of massive neutron, and the heavy hyperon Σ⁻. For the same density, the second produces much less pressure. The saturation density is indicated by an *arrow* [5]. Credit: Isaac Vidaña (CFisUC, Department of Physics, University of Coimbra)

For even higher densities $(2\text{–}3\rho_0)$, the very idea of "potential" fails (since it is a classical concept), and one must calculate the energy per nucleon using sophisticated techniques to obtain ϵ/n_B and then the pressure from

$$ P = n_B^2 \frac{\partial}{\partial n_B} \left(\frac{\epsilon}{n_B} \right) . $$

This difficulty means that, including more massive species such as the hyperons Λ and many others known from the laboratory, and depending on the type of treatment performed, there are substantial differences in the equations of state in the densest regime, which holds for over 90% of the mass of the star. This situation is illustrated in Fig. 6.14.

Finally, at the relevant densities, it may be that the degrees of freedom are *not* those we know from conventional nuclear Physics. There is strong evidence, both theoretical and experimental, of a phase transition where the nucleons release their fundamental constituents, quarks and gluons, under extreme conditions of temperature and pressure. While the RHIC and LHC experiments mainly explore the "hot" region of high temperature and low chemical potential (and consequently low density, since to a first approximation $\rho \propto \mu^4$ for relativistic matter), the domain of Astrophysics is a "cold" region close to the axis of the chemical potential μ, due to the fact that even the highest possible temperatures (tens of MeV) in supernova collapses are still very small when compared to the Fermi energy μ. For several decades there were improvements in the experiments to reach the quark–gluon plasma (QGP) region, and finally this phase was apparently detected in heavy ion collisions. We do not know to what extent these quarks are necessary to explain the interiors of neutron stars [27, 28].

6.3.3 Relativistic Stellar Structure Equations (TOV) and Neutron Stars

Just as Landau's argument in Chap. 3 allowed us to understand the existence of Chandrasekhar's mass without going into details, we can present an analogous discussion that allows us to visualize why there is a new branch of stable stars after the white dwarfs, and in passing, why there is no intermediate object between them. The argument is based on the behavior of the EoS studied previously, and on the concept of hydrostatic equilibrium which governs the star's structure. We have already seen that hydrostatic equilibrium is essentially a balance between gravitational forces and the forces associated with the pressure that supports the structure, that is, $P_G = P_M$, where we have used the name "gravitational pressure" P_G to designate the derivative of the gravitational energy with respect to volume, i.e., a quantity formally analogous to the matter pressure P_M. If we adopt the expression for the energy of a homogeneous sphere for the gravitational energy, viz., $E_G = -(3/5)(GM^2/R)$, and express it as a function of the volume $V = 4\pi R^3/3$, then assuming a constant mass, we have

$$E_G = -\frac{3}{5}\left(\frac{4\pi}{3}\right)^{1/3}\frac{GM^2}{V^{1/3}}. \tag{6.36}$$

It follows from (6.36) that $P_G = -\partial E_G/\partial V = C \times M^{2/3}\rho^{4/3}$, where C is a constant. We observe that this "gravitational pressure" has the same dependency on the density as an ultra-relativistic gas. Thus, $P_G = P_M$ is impossible unless the particles that supply the pressure become ultra-relativistic (!), since the pressures are then parallel. In the specific case of white dwarfs, the electron EoS starts as $P \propto \rho^{5/3}$, and then there are solutions for each constant mass (Fig. 6.15). However, when the total mass increases and the electrons become more relativistic, there are no solutions at any point since P_G and P_M have the same slope. Only when neutronization occurs and the neutrons themselves supply the pressure does the EoS change slope again, and equilibrium solutions then come back into existence. However, this happens for $\rho \geq \rho_0$, so the neutron star branch is established without any solution (star) in the middle of the density range. We see from this purely Newtonian analysis that we can infer the fundamental characteristics of this class of objects.

Analogously to the Newtonian equations of Chap. 3, an EoS of the form $P(\rho)$ is essential in order to integrate the TOV with the boundary conditions $M(r = 0) = 0$, $P(r = R) = 0$, together with the mass continuity equation $dM/dr = 4\pi\rho r^2$. In general, given the form (6.31) of the TOV, this requires numerical integration, but there are some cases where a complete analytical solution is possible. The set of exact

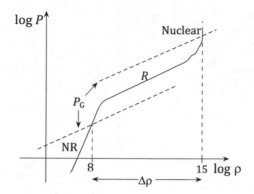

Fig. 6.15 The existence of equilibrium solutions is only possible when the slopes of the cold EoS and the "gravitational EoS" are different. The passage from the non-relativistic electron gas to the ultra-relativistic electron gas brings about the end of the stable model sequence. It is not until saturation density is reached that matter, now dominated by neutrons, is able to change the EOS slope again and stabilize neutron stars

solutions useful for modeling stars has been discussed by Lake [24], and contains the constant density solutions, Tolman V, and seven other cases.

A particular case of great interest leads to the so-called *Rhoades–Ruffini limit* [29], directly derived from a constant density approach. The reasoning behind this calculation is as follows: the greatest possible mass that can be supported by dense matter should occur when it is as "hard" as possible, that is, when $dP/d\rho = c^2$, and in fact a variational calculation confirms this expectation. If the EoS is considered known below a certain transition density ρ_T (described, for example, by the BBP EoS or similar), and also $dP/d\rho \geq 0$ locally to avoid any instability and collapse, then the maximum mass of the cold star sequence is

$$M_{RR} = 3.2 \left(\frac{4 \times 10^{14} \, \text{g cm}^{-3}}{\rho_T} \right)^{1/2} M_\odot \,. \tag{6.37}$$

This value can then be taken as an *absolute limit*, since there is no way to introduce any physical ingredient that makes the EoS violate causality, provided one assumes spherical symmetry, the absence of additional effects, and the validity of General Relativity. Obviously, the effects of rotation, for example, can slightly increase the value of the maximum, but only by about 20% or less. We will see that realistic neutron star models effectively keep the maximum mass of the sequence below this value. The general diagram of the mass-radius relations is shown in Fig. 6.16.

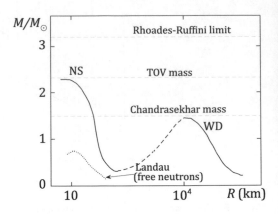

Fig. 6.16 Mass–radius relationship of white dwarfs and neutron stars on the same scale. The three limiting masses defined in the text are indicated with *dashed horizontal lines*

6.3.4 Stellar Models and Comparisons with Observations

Realistic neutron star models need to go far beyond this initial simplicity, as there are numerous developments in hadronic Physics to be incorporated, and they must also treat in detail the outer layers, where the magnetic field is anchored (see below) and the surface emits photons. A cross-section from a typical model is shown in Fig. 6.17.

In general, all models have an "atmosphere" from $\rho = 0$ to some $10^6 \, \mathrm{g \, cm^{-3}}$, composed of non-relativistic electrons and nuclei whose composition may depend on the fall of material in the supernova at the moment of birth. This characteristic is important since virtually all observable features including spectra are determined on this surface, where strong magnetic fields can affect the fundamental state, producing highly deformed nuclei and influencing these observed quantities [30].

Fig. 6.17 Profile of a typical neutron star model. *On the left*, the densities of each interphase, and *on the right*, the radius (in km) from the center

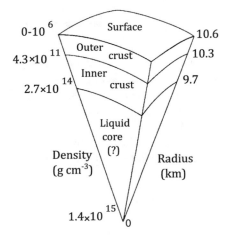

Above about 10^6 g cm^{-3}, and up to the value 4×10^{11} g cm^{-3}, the matter consists of relativistic electrons and a solid lattice of nuclei governed by Coulomb electrostatic forces. This is the so-called *outer crust*. We have already seen that, in addition to the neutron dripping point 4×10^{11} g cm^{-3}, such a gas coexists with the nuclear lattice, up to the nuclear saturation density $\rho_0 \sim 2.4 \times 10^{14}$ g cm^{-3}. The neutrons (without electric charge) are *superfluid* in this condition, i.e., they pair with each other and move without resistance.

Above the saturation density ρ_0 and up to the highest densities reached in the center, the uniform matter can at first be described with an EoS of the potential type, but then becomes uncertain, since the possibilities of the composition range from hyperon, meson, and/or quark condensates. This liquid lump remains one of the main unknowns of the structure, and contains more than 90% of the star's mass [27].

Star models are constructed by numerically integrating the structure equations for each chosen supranuclear EoS, and for each range of densities as the transition density points are reached, and this produces model sequences like those represented in Fig. 6.18. The models are more compact the higher their mass, and those that exceed the causal limit must be excluded (that is, they cannot enter the diagonal

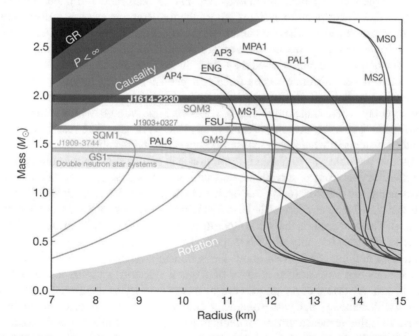

Fig. 6.18 Mass–radius diagram for neutron stars. *Horizontal stripes* indicate the highest measured masses, above about $M = 2M_\odot$. The stiffest equations of state are on the right, where the maximum masses are higher and the radius greater. The opposite happens with the softest equations of state on the left [31]. The three equations of state that contain quarks behave differently: the smaller masses have the smaller radii, because one assumes that the quarks are absolutely stable in the sense that, once released, they do not produce hadrons again. Some of the models in this figure are called *hybrids* (normal matter + quark core), because the quarks appear only at high pressure

gray range). This happens because, close to the maximum mass limit, the speed of sound in matter can exceed the speed of light, whence the calculation results become inconsistent. As the EoS is "harder" or "softer", i.e., it produces more or less pressure for a given energy density, the masses of the maximum mass models are higher or lower, respectively, while the corresponding radii follow the reverse trend. Thus one can construct the mass–radius diagram presented here.

To compare the models and the actual masses, the most widely used and successful method to obtain the latter has been the application of Kepler's third law in binary systems which contain at least one NS. The observations directly provide the so-called *mass function*:

$$f(M_1, M_2, i) = \frac{\left(M_2 \sin^2 i\right)^3}{\left(M_1 + M_2\right)^2} = \frac{T v_\parallel^3}{2\pi G}, \qquad (6.38)$$

since the binary period T and the projection of the orbital velocity of M_1 along the line of sight $v_\parallel$ can be measured. The angle of inclination i can sometimes be estimated, e.g., from the observation of eclipses or by combining data in other bands of the visible star, and then the masses M_1 and M_2 can be determined. The best results to date correspond to systems where the two components are neutron stars, but there are others where the companion M_1 is a white dwarf, an evolved star, or even a MS normal star. The measurement of orbital features plus the so-called *Shapiro delay*, the delay of the pulses when they "fall" in the potential well of the secondary, has produced very accurate values for a few pulsars (see, for instance, [31]) and is considered a "clean" observation because it is based purely on gravitation. The Shapiro delay is detectable only when the effect is large enough, which is why white dwarfs and a favorable geometry of the orbital plane are needed. The search continues for other systems in which such measurements would be feasible.

Another possible way to obtain information is to extract the effective temperature of a neutron star from observations, and with a distance evaluation, calculate the radius using the relation (4.1). But even though the spectra seem really thermal, with associated temperatures of 1–10 keV, the radii obtained are very small (around 3–5 km). The consensus is that the temperature obtained is not really the temperature of the whole star, but only of a "hot spot" (for example, polar caps) and that for this reason nothing can be said about the spherical structure, although there have been studies that have argued in favor of a quark star with a radius much smaller than 10 km. A variety of techniques are in progress to try to obtain the radius and mass simultaneously, but with somewhat conflicting results. A clear example of precise and reliable measurements of neutron star radii, essential to our understanding of cold, catalyzed matter beyond nuclear saturation density, have been provided recently by NASA's Neutron Star Interior Composition Explorer (NICER). High-quality data sets that have yielded measurements of the mass ($M = 1.44 \pm 0.15 M_\odot$) *and* radius ($R = 13^{+1.2}_{-1.0}$ km) of the 206 Hz pulsar PSR J0030+0451, and of the radius

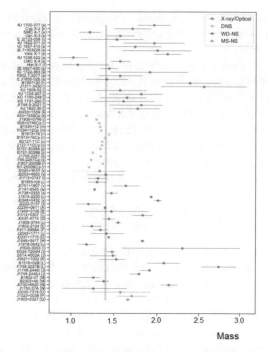

Fig. 6.19 Neutron star masses determined in binary systems. The best determinations are those corresponding to two neutron star systems, with more than one example of a "binary pulsar". Note that it can be proved that the distribution is *not* consistent with a single mass of $1.4M_\odot$, as stated in the literature before the 21st century. There are systems where the neutron star must have been added to $\Delta M \geq 0.3M_\odot$ and the distribution is at least bimodal, with one scale around $1.4M_\odot$ and another at around $1.8M_\odot$. A "third peak" (or rather, the "zeroth" peak) must be present at $1.25M_\odot$ with high significance in the current data, possibly associated with neutron star production by "light" progenitors (8–$10M_\odot$ in the MS) which develop degenerate O–Mg–Ne cores that collapse by electron capture. Compilation by L.S. Rocha [32]. See also [33]

($R = 13.7^{+2.6}_{-1.5}$ km) of the $M = 2.08 \pm 0.07M_\odot$, 346 Hz pulsar PSR J0740+6620 [34]. These numbers suggest that the radius is very similar for objects separated by about $0.5M_\odot$, a feature indicative of a stiff equation of state. The compilation of masses updated 2021 is seen in Fig. 6.19.

6.3.5 Pulsars and Other Neutron Stars

The first "real" (i.e., non-speculative) appearance of neutron stars in Astrophysics occurred in 1967, when researchers at the *Mullard Observatory* at Cambridge University [35] detected periodic temporal pulses from a cosmic source. A pulsing white dwarf was first considered as the most viable model, but the pulses were later associated by T. Gold and F. Pacini [36, 37] with a neutron star in rotation. This model readily gained support for its direct relationship with the old proposal by Baade and

Zwicky [20] about the compact remnants of supernovas, when the direct detection of the Crab pulsar was announced shortly afterwards, with an unsustainable period for a white dwarf pulse.

The basic idea of Pacini and Gold was to assign the radio pulses to the passage of an emission beam through the line of sight of the observer. In this way, the period of the pulses results directly from the rotation, but this mechanism also requires a magnetic field. Away from the object only the dipole should be important. A rotating dipole should lead to the torque exerted by the radiation that comes out and brakes the star. As the available energy source is the rotation itself, the losses should be equal to the change in the rotation energy, leading to the dynamic equation of a "spinning" magnetized star, with the rotation and magnetic axis at 90 degrees, so that the factor $sin^2\alpha = 1$:

$$I\Omega\dot{\Omega} = -\frac{2}{3c^2}B^2R^6\Omega^4 \, , \tag{6.39}$$

where the coefficient $2/3c^2$ corresponds to the radiation emitted by a rotating dipole in vacuum. Further research has shown that the electric field induced by this rotating dipole is so gigantic that a vacuum is not possible around the neutron star: electrons and protons are pulled from the surface by the induced electric field, and form a region around the pulsar where the dynamics of the particles is dominated by the magnetic field, which is thus called the *magnetosphere*.

After fifty years of research, this (classical) problem of the rotating dipole and the induced currents has not yet been fully solved. Although there are approximate solutions, the coefficient in (6.39) and other relevant quantities still cannot be calculated accurately [38]. In particular, we can calculate neither the detailed structure of the magnetic field, nor the flow of particles (or radiation) escaping along the open lines of the magnetosphere in the form of a relativistic wind, clearly visible in X-rays (Fig. 6.20).

However, assuming that only the dipole emission contributes to braking the rotation, and that the magnetic field does not change throughout the life of the star, we can integrate (6.39) with respect to time to give

$$t = -\frac{\Omega}{(n-1)\dot{\Omega}}\left(1 - \frac{\Omega^{n-1}}{\Omega_i^{n-1}}\right) \, , \tag{6.40}$$

where Ω_i is the initial rotation speed of the pulsar and n is the so-called *braking index*, which measures the braking of the object as the radiation flows. If the pulsar was born rotating much faster than today, the term in brackets is approximately unity and we can define the *characteristic age* $\tau = P/2\dot{P}$ as the typical timescale for the rotation to decrease. A pure dipole leads to the value $n = 3$, but the definition $n = \Omega|\ddot{\Omega}|/\dot{\Omega}^2$ can be formulated as a directly observable quantity if the speed of rotation and its first two derivatives are determined [4]. Table 6.2 displays the values for 6 pulsars where it has been possible to determine the 3 quantities (especially the tiny second derivative $\ddot{\Omega}$), and all of them differ from the expected value, in some cases substantially. This difference indicates that the energy loss is not purely from dipole radiation, and that other factors are involved. This discrepancy is not new. The Crab pulsar, for example,

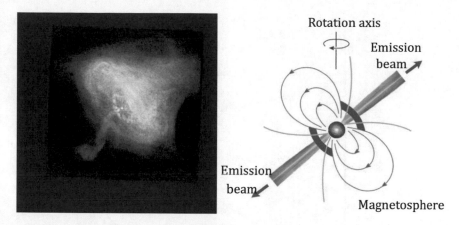

Fig. 6.20 *Left*: X-ray image of the Crab pulsar, oriented in a similar way to the schematic figure in the sky plane. Credit: NASA/CXC/SAO/F. Seward et al. *Right*: Schematic of a pulsar showing the co-rotation of the particles with the neutron star right out to the light cylinder (line indicated with the *arrow*) and the emission in the direction of the observer. Credit: NRAO

Table 6.2 Some values of the braking index measured so far

Pulsar	P [s]	$\dot{P}$ [10^{-13} s/s]	Braking index n
PSR B0833-45 (Vela)	0.089	1.25	1.4 (2)
PSR B0540-69	0.050	4.79	2.14 (9)
PSR J1846-0258	0.324	71.0	2.16 (3)
PSR B0531+21 (Crab)	0.033	4.23	2.51 (1)
PSR J1119-6127	0.408	40.2	2.91 (5)
PSR B1509-58	0.151	15.4	2.839 (3)
PSR J1734-3333	1.17	22.8	0.9 (2)
PSR J1833-1034	0.062	2.02	1.8569 (6)
PSR J1640-4631	0.207	9.72	3.15 (3)

emits around 10^{31} erg in pulsed radiation, while $I\Omega\dot{\Omega} > 6 \times 10^{38}$ erg, i.e., although it is assumed to be the main emission, the dipole radiation takes no more than a small fraction of the total energy. The presence of other energy fluxes is necessary, and in particular there is the unequivocal detection of a particle flow (wind), which collides with the circumstellar material to produce X-rays, as observed in Fig. 6.20.

Despite these uncertainties and reservations, the dynamic equation (6.39) is widely used in the form

$$B = 10^{13} \sqrt{\left(\frac{P}{1\,\text{s}}\right)\left(\frac{\dot{P}}{10^{-13}\,\text{s s}^{-1}}\right)}\ \text{G}\,, \tag{6.41}$$

to obtain an estimate of the magnetic field B as a function of the observables P and $\dot{P}$. This results from the direct inversion of (6.39), and from the hypothesis that

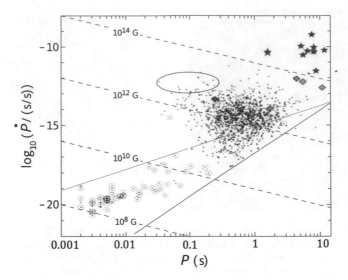

Fig. 6.21 The log $\dot{P}$–log P diagram for pulsars and similar objects [45, Fig. 2]. Lines $B =$ constant are shown explicitly. Ordinary pulsars are grouped in the central region. Other neutron stars populate this fundamental diagram: the magnetars in the *upper right corner* (AXP-SGR, in *red*) and the millisecond pulsars (recycled) in the *lower left corner*. The *blue diagonal line* is called the "death line", and marks where the pulsars are too slow or too demagnetized to produce emission. Note the set of pulsars associated with a supernova remnant, marked with an *ellipse*. Credit: V.M. Kaspi

$\sin^2 \alpha \sim 1$. With (6.41) and the characteristic age τ of (6.40), we can calculate the paths of the pulsars in the log $\dot{P}$–log P diagram, which yields straight lines for each set $B =$ constant (Fig. 6.21). Although the field was expected to decay due to Ohmic dissipation in the crust, the existence of a large group of pulsars with very high values of τ and very intense fields has somewhat discredited the field decay theory, but not completely ruled it out.

The presence of pulsars detected only in the high-energy bands and not in radio deserves a comment. While it has been suggested that pulsed radio emission is *incoherent* (although the standard picture points to a coherent nature), optical, X, and γ emissions are surely *coherent*, and thus result in a proportional density of the emitting particles. There are several mechanisms that can explain these emissions. The most interesting is the presence of thermal radiation from the surface residue of the content of the birth, since it provides information about the cooling processes and therefore the state of the interior. The charged particles in the magnetosphere and winds can produce incoherent emission. In particular, there are a number of important detections in the γ bands that should help to explain the classification of pulsars, but the content of the emission at the higher energies remains open.

In relation to the "other" types of neutron stars shown in Fig. 6.21, the most extreme is the *magnetar*. In the decades following their discovery, some sources were identified which, besides presenting an emission in X-rays much greater than $I \Omega \dot{\Omega}$ (that is, the energy of the observed emission cannot be obtained from their rotational

Fig. 6.22 The SGR 1806-20 magnetar outburst of December 27, 2004, observed by the RHESSI probe [42]. The energy in gamma rays is equivalent to a million years of solar emission (!) released in "hard" photons in about 300 s. It is better not to be anywhere near the source when this happens. Credit: Hurley et al.

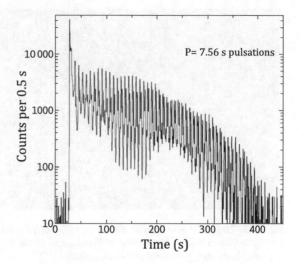

energy, which is insufficient), have long periods, longer than 1 s, and values of the derivatives well above those of ordinary pulsars (of order 10^{-10} s s^{-1}). Thus, (6.41) shows that their magnetic fields should be 10^{14}–10^{15} G, and for this reason they became known as *magnetars*. This class of sources is also observed in γ rays, often in the form of bursts and intense activity (Fig. 6.22). The idea of the magnetar model is that the sudden dissipation of magnetic energy is responsible for this phenomenon. The model was applied to the group called *soft-gamma repeaters* and *anomalous X-ray pulsars* (SGR-AXP), considered to be different manifestations of neutron stars with extreme magnetic fields [39]. However, there are recent detections such as SGR 0418+5729, with an estimated magnetic field of 7.5×10^{12} G, much lower than the others, and they are sometimes detected in radio. This requires a rethink of the magnetar scenario, since with such a low-field it may not be possible to extract enough rotational energy to explain the observed X-ray emission.

The association of pulsars with supernova remnants is today a well-established idea, although it has been confirmed in less than 20% of the more than 200 SNR known in the galaxy. These associations are an important problem. Most remnants should *not* be associated with a pulsar, since type Ia supernovas do not produce pulsars and there should also be cases where the product was a stellar black hole (Chap. 5). Another important factor is that pulsars are born (on average) with high proper motion due to the birth process, and reach speeds of the order of 400 km s^{-1}. Thus, pulsars often "punch" the edge of the young remnant and escape. Finally, the emission beam can point away from the Earth (in 70–80% of cases), and identification of the SNR itself becomes almost impossible after more than 10^5 yr (Chap. 5). Several remnants have been associated with magnetars, but there are major problems in confirming these associations. The attempt to associate magnetars with *massive star clusters* that may have been their progenitors also presents problems. Although it has been suggested that a number of cases indicate progenitors of more than $40 M_\odot$ (raising

Fig. 6.23 The globular cluster 47 Tuc in the optical (*left*) and X-ray (*right*). Some of the pulsars belonging to this system are marked with *red stars*. In the X-ray image we can see these pulsars and other high-energy systems that emit intensely [43]

the question as to why such massive stars did *not* form black holes), there is at least one case in which the cluster still has stars of mass of the order of $17 M_\odot$ on the MS, casting doubts about the higher masses [40] (unless NS/BH formation is actually an intermittent function of the mass of the progenitor, going from one to the other and back).

Finally, Fig. 6.23 also shows the *millisecond* pulsars, a class that includes the fastest rotating object known today, PSR J1748-2446ad in the Terzan 5 cluster, with $P = 1.4$ ms (or a frequency of 716 Hz). This object is another representative of a class detected more than 30 years ago, and which contains an important number of pulsars in globular clusters. As the clusters have not suffered many collapse supernovas, an alternative channel of formation by the addition of matter on top of a white dwarf has been postulated, known as *accretion induced collapse (AIC)*, where electron capture should be faster than carbon ignition (see Chap. 5). It is believed that the ultrashort periods of millisecond pulsars may be due to *recycling* [41], that is, to the process in which the addition of the "normal" companion transfers angular momentum and thus accelerates the rotation. Conditions for the existence of these systems are highly favorable in clusters, and indeed most of the pulsars detected in them are in the millisecond class, although there may be some with the original rotation, without having undergone recycling.

The latest discovery of an entirely new class of compact stars is the so-called *Rotating Radio Transient Sources* (RRATS), which emit sporadic radio pulses in phase, separated by several hours, and may constitute the dominant population of the disc, given their characteristic ages and detection difficulties. In other words, RRATs would be the overwhelming, almost totally silent population of neutron stars that will continue to be a frontier research topic for decades to come, even more so after the first detection of the fusion of two into gravitational waves and radiation across the entire electromagnetic spectrum (Chap. 10).

Fig. 6.24 Pioneers of the black hole idea. *On the left*, John Michell, member of the British Royal Geological Society. *On the right*, Pierre Simon, the Marquis of Laplace

6.4 Physics and Observational Manifestations of Black Holes

6.4.1 Birth of the Black Hole Concept

The long history of the black hole idea has two illustrious precursors in the late 18th century (!). Within a few years of each other, Englishman J. Michell and Frenchman Pierre-Simon de Laplace (Fig. 6.24) discussed the possibility of *dark stars* based on Newtonian ideas about the escape velocity of corpuscles of light from the star surface. Note that these arguments are based on the Newtonian concept of the corpuscular nature of light, otherwise there would be no way Newton's force of gravity could attract it. But despite this, Michell and Laplace's reasoning opened the way to the modern study of black holes and deserves a discussion [44].

As is well known, one uses energy conservation to obtain the critical condition for a particle to escape from the surface of a body of mass M and radius R :

$$\frac{1}{2}mv^2 = \frac{GMm}{R} \ . \tag{6.42}$$

If we set $v = c$, applicable to light in this Newtonian corpuscular approach, we find that when the radius reaches the critical value $R = 2GM/c^2$, the gravitational field will not allow the particle to escape. Thus, the compact object that reaches this condition will appear to be "dark", and invisible to outside observers.

Two centuries later, when the General Theory of Relativity was formulated, formal solutions were discovered in which, instead of considering a physical surface for the emission of light, there is an imaginary surface called the *event horizon* from which no point outside can receive any signal from the interior. The exterior and interior of the horizon are causally disconnected. This is a consequence of the strong curvature induced by the concentration of mass. Moreover, right in the center there is

Fig. 6.25 Two-dimensional diagram of a black hole. The singularity is covered by the horizon, which separates the "interior" from the rest of the Universe

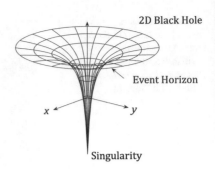

a *singularity* which cannot be described by the classical equations of GR and which is "hidden" inside the horizon. This basic solution, without adding angular momentum, electric charge, or other ingredients, is the so-called *Schwarzschild solution*, obtained by the German physicist of the same name while fighting in the trenches of World War I on the Russian front (students should note when claiming they didn't have the right conditions to do their homework). A schematic view of Schwarzschild's black hole structure is shown in Fig. 6.25.

With the physical idea of *compactness* as the cause of the formation of a black hole we can go beyond and build a general diagram which contains all the possible black holes in the Universe, including the Universe itself. This is known as *Carter's diagram* (see Fig. 6.26). We have already seen that the configuration radius has a

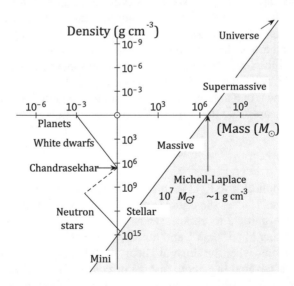

Fig. 6.26 Carter diagram [47]. The *grey region* is the black hole domain, that is, mass compressed beyond the density inversely proportional to R_S^3 forms the so-called *event horizon* (see text). Note that as the mass increases, the effective density *decreases*. The Michell–Laplace black hole is indicated, since they imagined that the density was kept constant at $1 \, \mathrm{g \, cm^{-3}}$, as in the Sun. The Universe itself could enter its Schwarzschild radius and form a black hole without us noticing

critical value for $R = 2GM/c^2 \equiv R_S$, which by chance is exactly the value of the so-called Schwarzschild radius obtained rigorously in General Relativity, and which marks the position of the horizon. We can gain some understanding of R_S by noting that the bracketed term in the denominator of (6.31) is $1 - 2GM/rc^2$, and while $2GM/rc^2 \approx 0.1$ in neutron stars, black holes with $R_S = 2GM/c^2$ make this term zero, and the TOV description no longer makes sense. Physically, we can think of neutron stars as tightly "packed", but the pressure still resists gravitation. On the other hand, black holes reach a critical level "packing" and all the matter disappears behind the horizon. Thus, we will not need to impose any equation of state, since black holes result in "pure gravitation".

Carter's diagram [47] has on the vertical axis the "density" of the objects. This may seem a bit strange, since we have just said that matter is not present and has collapsed within the Schwarzschild radius. However, it is always possible to define a formal density $\rho_{BH} = 3M/4\pi R_S^3$ which, combined with the definition $R_S = 2GM/c^2$, implies that $\rho_{BH} \propto 1/M^2$. Black holes with mass much greater than $10^6 M_\odot$ are called *supermassive*. There is one in the center of our galaxy, in Sgr A*. They are much less dense than water, while a miniature black hole with mass much less than $M_\odot$ is much denser than a neutron star.

Until the second half of the 20th century, isolated black holes were not expected to be very interesting. On the other hand, those that accrete matter from a companion (stellar case) or the circumstellar medium (supermassive case) are of great interest, as we will see below. But even those that do not have a companion were studied and a very interesting result was obtained: the very intense gravitational field near the horizon has the property of causing *vacuum fluctuations* (see Fig. 1.2), and one of the particles in the resulting particle–antiparticle pairs is sometimes absorbed behind the horizon, while the "orphan" companion escapes the system to infinity. Adding up all the contributions, the total spectrum is thermal, with a temperature (called the Hawking temperature) inversely proportional to the mass of the black hole (Fig. 6.27). This is a heuristic justification of the *Hawking radiation*, expected theoretically from an otherwise inert black hole.

Using the discussion in Chap. 1, we can substantiate these statements and obtain the Hawking temperature. The emission of a photon near the horizon implies that the uncertainty in its position is of the order of the radius R_S, that is,

$$\Delta x \approx R_S = \frac{2GM}{c^2} . \tag{6.43}$$

From this we can immediately calculate the uncertainty Δp in the momentum as

$$\Delta p \approx \frac{\hbar}{\Delta x} = \frac{\hbar c^2}{4GM} . \tag{6.44}$$

The typical photon energy is thus $E_\gamma = \Delta pc = \hbar c^3/4GM$. Associating a temperature T_H with this characteristic energy and introducing a numerical factor 4π that cannot be easily obtained with this simple estimate, we find

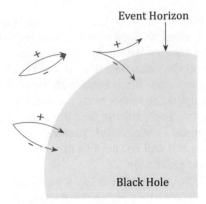

Fig. 6.27 Vacuum fluctuations as the source of Hawking radiation. Most of the time the particle–antiparticle pair will annihilate without great consequences, as happens in the absence of gravitation. Sometimes the pair will fall behind the horizon, but it also happens that only one of the particles is captured while the other escapes. When one sums over the latter particles, one finds that black holes emit radiation with the spectrum of a black body [48]

$$T_{\mathrm{H}} = \frac{\hbar c^3}{8\pi GM} . \tag{6.45}$$

That is, the temperature of the black body emission goes as the reciprocal of the black hole mass. Numerically, we can express this in K as

$$T_{\mathrm{H}} = 10^{-7} \frac{M_\odot}{M} \, \mathrm{K} . \tag{6.46}$$

From (6.46) it is evident that Hawking's radiation is very weak and totally unobservable, unless the evaporating black hole is close to total disappearance ($M \to 0$). One of Hawking's suggestions was to monitor very brief gamma ray bursts as a sign of the end of evaporation.

However, there is another context in which Hawking radiation can be very important: for the fate of primordial black holes, produced very early in the history of the Universe. There are several possible mechanisms for black holes to form. In the simplest (collapse of large density fluctuations), the candidate mechanism must be able to create fluctuations $\Delta\rho/\rho \geq 1/3$ in the nearly homogeneous primordial matter. The fluctuations detected today on a variety of scales by monitoring the cosmic background radiation (i.e., temperature fluctuations that reflect fluctuations in density at the time) are of the order of 10^{-5}, and it is possible that there are large enough fluctuations, as required, but that they remain "hidden" at small scales. Regardless of this, we can still discuss what the evaporation of black holes would be like in the cosmological context. The Hawking emission, identified with that of a black body with temperature T_{H}, implies a loss of energy from the black hole at a rate

$$\dot{E} = -4\pi R_{\mathrm{S}}^2 T_{\mathrm{H}}^4 . \tag{6.47}$$

As we have seen before, $R_S \propto M$ and $T_H \propto 1/M$, and the rate of loss of mass (energy) of the black hole is then

$$\frac{dM}{dt} = -\frac{A}{M^2} . \tag{6.48}$$

On the other hand, primordial black holes are immersed in a very energetic environment and absorb particles and radiation from the environment. A complete calculation of the cross-section (which takes into account the fact that black hole gravitation increases the geometric section) results in $\sigma = (27\pi/4)R_S^2$, larger than the geometric cross-section because of the attractive effect of gravitation. This absorption effect can only be important in the so-called *radiation-dominated era*, because when matter begins to dominate the expansion, the flux of energy falling into the black hole becomes insignificant. Using the cross-section and the fact that $F \propto c\rho_{rad} \propto T_{rad}^4$, and assuming that the black holes constitute a dilute gas and do not contribute much to the dynamics of the Universe, we find

$$\frac{dM}{dt} = -\frac{A}{M^2} + BM^2 T_{rad}^4 , \tag{6.49}$$

for the evolution of the hole mass with time, where A and B are calculable constants depending on the time considered and $T_{rad} = T_{rad}(t)$ is the temperature function derived from Friedmann's equations for the evolution of the Universe. Note that, keeping only the first Hawking term, we can integrate to obtain the initial mass of a black hole that would be evaporating today, corresponding to a point object, but with a mass similar to that of an asteroid:

$$M \equiv M_H \approx 5 \times 10^{14} \, g . \tag{6.50}$$

Inclusion of the absorbed energy leads us to define a curve called the *critical mass* $M_C(t)$ that separates the regions where the black hole absorbs energy or evaporates (Fig. 6.28). This stems from the condition $dM/dt = 0$ in (6.49) [49]. The critical mass is a property of the environment and, in the radiation-dominated era, has the value

$$M_C(t) = 10^{26} \frac{T_0}{T_{rad}(t)} \, g , \tag{6.51}$$

where T_0 is the cosmic temperature at which the primordial black holes form. These developments continue today, and serve, for example, to identify which times and mechanisms would have given rise to primordial black holes that would evaporate today or contribute to the observed IR, radio, etc., radiation backgrounds. A complete overview can be found in [50].

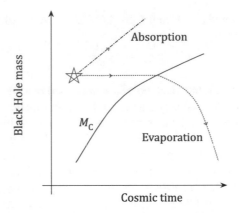

Fig. 6.28 Evolution of a black hole in the early Universe. If formed above the critical mass at the time, the black hole has almost constant mass (horizontal path) until it reaches the instantaneous value of M_C in a future time, and will start evaporating from this instant. Zel'dovich and others were concerned about the possibility of black holes growing explosively, thus swallowing a substantial fraction of the Universe, something that does not happen in practice. The delay before the black holes begin to evaporate may be substantial [51]

6.4.2 What Do We Observe from Black Holes?

Until the middle of the 20th century, research on black holes had a very different character from today's. Mathematical and formal aspects of the solutions, alternative theories, imaginary experiments, and other problems of this type were discussed. However, the observation of black holes were never considered, and neither was the modern idea of the production of black holes in massive star collapses, while the concept of a supermassive black hole appearing on the right of Carter's diagram was never formulated in connection with any observation.

What really boosted the empirical study of black holes was the discovery of quasars in the 1960s (more details in Chap. 8), since the energy source pointed to a highly efficient emission mechanism, and giant mass black holes were then seriously considered for the first time. It was around this time that the physicist John Archibald Wheeler pulled off a major "publicity stunt" when, in a 1967 lecture, he called solutions with an event horizon *black holes* (although there were antecedents to this name), rather than "frozen stars" as they had been called in Russia. The name totally changed our view of these objects, and the consideration of quasars brought them into the realm of reality, while they remained interesting for mathematics.

A particular aspect of the study of "real" black holes is the behavior of light when emitted by sources near the horizon, since we know that curvature produces significant distortions with respect to the usual propagation. Simulated images of the effects of a black hole on light have been produced since then (Fig. 6.29), but not yet directly observed.

Explorations of the distortion of light by the gravitational field has prompted an ambitious initiative to image the neighborhood of a black hole's event horizon. It

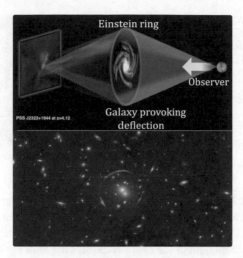

Fig. 6.29 Gravitational lensing: the distortion of images by gravitation. *Upper*: Deflection of light by the gravitational field of a large object, such as a galaxy or cluster of galaxies, produces multiple images of a source depending on the exact geometry (r_{LS}, r_L, etc.). Credit: Bill Saxton, NRAO/AUI/NSF. *Lower*: Example of a nearly complete arc image of a galaxy with a background distorted by a cluster situated in front of it. The passage of light near the event horizon of a black hole is an extreme case of this phenomenon, already demonstrated in other extragalactic systems. Credit: ESA/Hubble & NASA

Fig. 6.30 *Left*: The Event Horizon Telescope showing some of the instruments [52]. *Right*: The aim is to produce images with μarcsec resolution, enough to "see" the supermassive black hole in Sgr A* at the center of our galaxy. Credit: EHT Collaboration

should be possible to observe image distortions directly and even compare various possibilities that arise from different theories of gravitation against the prediction of GR. It is clear that this requires a huge angular resolution, since a black hole occupies little more than a point. These observations are conducted by the network called the *Event Horizon Telescope*, using the largest possible baseline, on the order of the Earth's diameter (Fig. 6.30). An angular resolution of order the μarcsec is required to produce images of the type shown.

Fig. 6.31 Image of the
central black hole in M87
showing the ring of radiation
produced by the bend in the
gravitational field and the
central "shadow" [52].
Credit: EHT Collaboration

In April 2019 the EHT collaboration announced its first concrete result [53], i.e.,
the first image of a black hole horizon, for the black hole at the center of the M87
galaxy (Fig. 6.31). The image shows a bright ring caused by the distortion of light
through the gravitational field of the black hole, and the dark region that is produced
by capturing photons through the horizon, also known as the "shadow" of the black
hole. The horizon is in fact much smaller than the dark region, but the "shadow" is at
the limit of what can be imaged. Comparison with numerical simulations indicates
that General Relativity reproduces the measured distortion well, and there is little or
no evidence for an alternative theory, a result that should be confirmed in the future
for this and other cases when the resolution is improved and can constrain alternative
theories more tightly.

We now describe the two categories of black holes for which we have a certain
amount of information (the first category, already discussed, comprised the primordial black holes, but it remains unconfirmed). The closest and most abundant category
are the black holes produced by high-mass stars, possibly starting out with masses
of $25M_\odot$ or more in a prompt fashion, and alternating with neutron stars below this
range. Although this lower limit is uncertain, there is a strong consensus that, above
a certain threshold value, the iron core and the resulting explosion dynamics will not
be able to produce a neutron star, whence the result will be a black hole of stellar
mass. From an observational point of view, the very nature of black holes suggests
that there will only be possibilities of observing them successfully when they are
members of binary systems. These may or may not be in states of accretion, depending on the evolution of the companion and the orbit. But for those systems in which
it is possible to determine a mass for the compact object, the Rhoades–Ruffini limit
in (6.37) provides a very reliable test to distinguish between a neutron star and a
black hole. Figure 6.32 shows a diagram of a binary sample where this determination was possible. The objects listed are black hole candidates because they exceed
the Rhoades–Ruffini limit (explicitly indicated). The existence of small-mass black
holes and massive neutron stars, thereby eliminating the so-called *mass gap*, is an
important topic in compact object Astrophysics.

One of the notable examples of the identification of binary systems in Fig. 6.32 is
the extragalactic X-ray binary M33 X7 (Fig. 6.33). Every 3.45 days the companion, a
star of mass around $70M_\odot$, is eclipsed by the disk that passes through the line of sight.

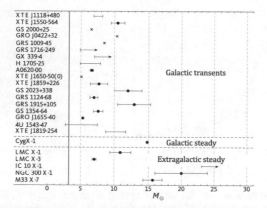

Fig. 6.32 Black hole candidates in binary systems. All systems have inferred masses higher than the Rhoades–Ruffini limit (*blue vertical line*), allowing their identification. Note the absence of low-mass candidates, and the maximum value of order 15–20$M_{\odot}$ for galactic objects. From [54]

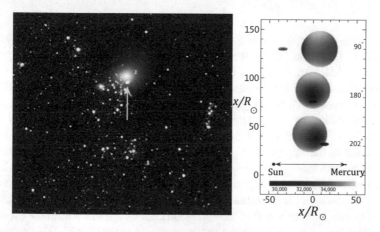

Fig. 6.33 The extragalactic binary M33 X-7 in X-rays. *On the left*, normal binary emission. *On the right*, the eclipse through the disk (it does not fully zero because something from the X radiation spreads out of the disc). Note that, to reproduce the light curve, the semi-axis of the orbit is less than 1/2 that of the orbit of Mercury [55]. Credit: NASA/CXC/CFA/P. Plucinsky et al. (X-ray), NASA/STScI/SDSU/J. Orosz et al. (optical)/Science Photo Library

This makes it possible to determine with great accuracy the inclination sin i in (6.38), and with it the mass of the primary dark object, giving a result $15.65 \pm 1.45 M_{\odot}$. Thus, this result identifies the primary as a black hole. This corresponds to the last line in Fig. 6.32 and is one of the most reliable determinations, thanks to the presence of eclipses.

There are other ways to determine the presence of black holes in systems that go beyond the use of Kepler's third law. For example, several X-ray binaries have X-outs, where the "hard" photon count rises very quickly and decreases in a few

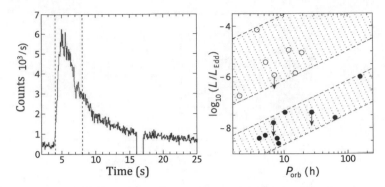

Fig. 6.34 X-ray outbursts as a test for the presence of black holes. *Left*: Typical burst, where the count increases by at least a factor of 10 and returns to the initial state after about 2 minutes. *Right*: $L-P_{orb}$ diagram for a set of sources. The separation into two groups of different luminosity is evident, and happens independently from P_{orb}. Those at the bottom are identified as black holes [57, Fig. 3]. © AAS. Reproduced with permission

minutes (Fig. 6.34 left). The most widely accepted interpretation is that hydrogen accumulates on the surface of the star until it reaches the density and temperature of thermonuclear ignition. X-ray binaries can be analyzed in various ways, but one revealing diagram is that of luminosity vs. orbital period. When placed in this plane, a *gap* is observed between two groups, one more luminous and one much less luminous in the stationary states. The key observation is that only the brightest group (Fig. 6.34 top right) presents X-outbursts, while the least bright group never does. This has led to the interpretation that, since the outbursts need the existence of a surface to accumulate hydrogen, the brightest group contain neutron stars, while the less luminous group was composed of black holes that had no surface and no outbursts could occur. This hypothesis would also explain the difference in stationary luminosity: the innermost part of the accretion disk would fall within the event horizon and that is precisely where most of the energy is radiated [56]. We see here how tests can be proposed to discover the presence of black holes in real stellar systems.

Besides the class of stellar mass black holes, we now have evidence for the presence in the Universe of supermassive black holes, with masses above $10^6 M_\odot$. This class was not considered before the 1960s, and it was precisely the discovery of quasars that led to its study. These developments will be analyzed in Chap. 8, but we will discuss one of these cosmic monsters briefly here, of particular interest because it is located in the center of our own galaxy.

The study of the center of our galaxy is not at all easy, since the region is strongly obscured by dust and gas. There are some "windows" of low extinction in which observation is somewhat easier, and of course infrared wavelengths and radio waves can be used for this research. With the accumulation of information over time, it became clear that the central parsec contains a very interesting stellar population, in addition to compact objects and supernova remnants. Studies of the motion of the

Fig. 6.35 The center of the Milky Way (Sgr A*), indicated by the *arrow*, and several star in orbits around it. This image was created by Prof. Andrea Ghez and her research team at UCLA and is from a data set obtained with the W.M. Keck Telescopes. © KECK/UCLA Galactic Center Group

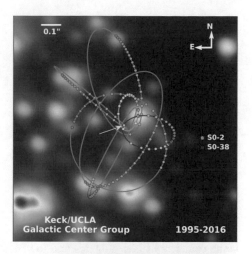

Fig. 6.36 Example of binary black holes in an outer galaxy (NGC 6240). The history of formation is still quite a mystery (see Chap. 8). Credit: NASA/CXC/MPE/S. Komossa et al.

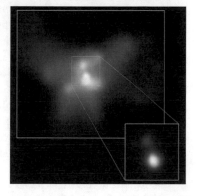

most central stars attested to the presence of a compact supermassive object using the method from Kepler's third law already explained. Figure 6.35 shows the orbits of two particularly interesting stars (S0-2 and SO-102). These have relatively short periods and were well determined after more than 15 years of observations [58]. The direct application of Kepler's law shows that there is an object of mass calculated to be $3.5 \times 10^6 M_\odot$ sitting almost at the focus of the ellipses (marked with an arrow). There is no signal at any wavelength that reveals the presence of this object, and for this reason it is believed to be a giant black hole. There are other proposals that are more exotic than the black hole, but as there are known to be supermassive black holes in a huge number of galaxies, there is a strong consensus in favor of this hypothesis based on kinematic observations.

These objects in galactic centers are often active, giving rise to the so-called AGNs (Chap. 8). But of course the center of the Milky Way is not an active galactic nucleus. This is because the fall of matter onto the central object is only sporadic, in contrast with its cosmological relatives. However, it has been possible to directly imagine

Fig. 6.37 Observed profile of the K_α iron line in NGC 4151 [59]. The distortion (asymmetry) is evident, but there are only a handful of such systems

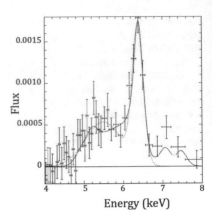

some centers of outer galaxies and to check that black holes are present, in some cases even forming multiple systems (Fig. 6.36).

Other cases where effects directly associated with the presence of a supermassive object have been observed include the so-called K_α iron line, identified with the fluorescence of the K-layer. This emission line is present in X-rays, with an energy of 6.4 keV. Its profile has been studied and is well known. But in at least two extragalactic systems, the line is strongly asymmetrically distorted (Fig. 6.37). The interpretation is that the emitting material is accreting onto a large mass black hole, so that the gravitational "pull" towards the center appears clearly in the spectrum. It is important to point out that, despite several campaigns to look for more examples, these are scarce. It is not clear why there should be this rarity, since the accretion phenomenon is very common and must be associated with galactic activity (Chap. 8).

Finally, we will discuss a class of sources discovered in 1994 by L.F. Rodríguez (UNAM, México) and F. Mirabel (IAFE, Argentina) that brought a completely different perspective on the nearest black holes. Mirabel and Rodríguez observed Cyg X-1, an object that showed the presence of relativistic jets and radio lobes similar to those observed in the AGNs (Fig. 6.38 left), but on a much smaller scale. The

Fig. 6.38 *Left*: Image of Cyg X-1 with its jets and radio lobes near the center of our galaxy. *Right*: Superluminal motion of the GRS 1915 jets. The interval between the first and the last image is less than 1 month, and the apparent distance covered by the material is about 8000 A.U. [60]

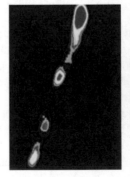
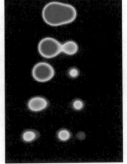

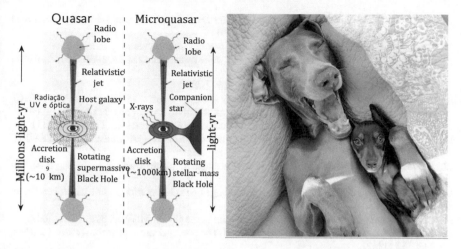

Fig. 6.39   Quasar and microquasar. *Left*: Comparison identifying the main elements of each system. Credit: I.F. Mirabel. *Right*: Biological analogy (not to scale) that emphasizes the identity of the structure and processes, but on very different scales

distance inferred for Cyg X-1 is about 8 kpc. Soon after, the same researchers were able to show that a second source (GRS 1915) has so-called superluminal jet motion (Fig. 6.38 right), in which the structures of the jet seem to move away at speeds greater than c due to a projection effect [60]. It was thus demonstrated that the jets were relativistic and that, in general, objects of stellar mass (GRS 1915 contains a black hole of estimated mass around $33 M_\odot$) behave to a large extent like their gigantic "cousins" in the AGNs. Hence the name *microquasars*, with which they are still known today.

Figure 6.39 compares the morphologies of quasars and microquasars. While in the case of AGNs the accretion is thought to be "fossil", from the surrounding environment, microquasars are fed by a donor companion. It is important to point out that, due to the difference in spatial and temporal scales, we can see phenomena that would otherwise be very slow or distant. For example, the superluminal expansion analogous to Fig. 6.38 takes many years in the case of an AGN. Thus we have nearby systems that behave very much like the more remote ones, and hence the possibility of learning more about the central objects that are identified with black holes.

References

1. E.E. Salpeter, The luminosity function and Stellar Evolution. Astrophys. J. **121**, 161 (1955)
2. https://www.wikiwand.com/en/Initial_mass_function
3. S.E. Woosley, A. Heger, T.A. Weaver, The evolution and explosion of massive stars. Rev. Mod. Phys. **74**, 1015 (2002)

4. S. Shapiro, S.L. Teukolsky, *Black Holes, White Dwarfs and Neutron Stars: The Physics of Compact Objects* (Wiley-VCH, Weinheim, 1991)
5. https://compstar.uni-frankfurt.de/outreach/short-articles/the-hyperon-puzzle/
6. J.B. Holberg, The discovery of the existence of white dwarf stars: 1862 to 1930. J. Hist. Astron. **40**, 137 (2009)
7. R.H. Fowler, On dense matter. MNRAS **87**, 114 (1926)
8. S. Chandrasekhar, *An Introduction to the Study of Stellar Structure* (Dover, New York, 2010)
9. R.F. Tooper, Stability of massive stars in General Relativity. Astrophys. J. **140**, 434 (1964)
10. S.O. Kepler et al., White dwarf stars. Int. J. Mod. Phys. Conf. Series **45**, 1760023 (2017)
11. S.O. Kepler, D. Koester, G. Ourique, A white dwarf with an oxygen atmosphere. Science **352**, 6281 (2016)
12. I.-S. Suh, G. Mathews, Mass-radius relation for magnetic white dwarfs. Astrophys. J. **530**, 949 (2000). https://doi.org/10.1086/308403
13. E.E. Giorgi et al., NGC 2571: An intermediate-age open cluster with a white dwarf candidate. Astron. Astrophys. **381**, 884 (2002)
14. S.O. Kepler et al., White dwarf mass distribution in the SDSS. MNRAS **375**, 1315 (2007)
15. L. Mestel, On the theory of white dwarf stars. I. The energy sources of white dwarfs. MNRAS **112**, 583 (1952)
16. H.B. Richer et al., White dwarfs in globular clusters: Hubble Space Telescope observations of M4. Astrophys. J. **484**, 741 (1997). https://doi.org/10.1086/304379
17. A. Kanaan et al., Whole Earth Telescope observations of BPM 37093: A seismological test of crystallization theory in white dwarfs. Astron. Astrophys. **432**, 219 (2005)
18. L.D. Landau, On the theory of stars. Phys. Z. Sowjet. **1**, 285 (1932)
19. D.G. Yakovlev et al., Lev Landau and the conception of neutron stars. Phys. Uspekhi **56**, 289 (2013)
20. W. Baade and F. Zwicky, op. cit. Chap. 5
21. R. Oppenheimer, G. Volkoff, On massive neutron cores. Phys. Rev. **55**, 374 (1939)
22. B.F. Schutz, *A First Course in General Relativity* (Cambridge University Press, Cambridge, UK, 2009)
23. R.C. Tolman, *Relativity, Thermodynamics and Cosmology* (Dover, New York, 2011)
24. M.S. Delgaty, K. Lake, Physical acceptability of isolated, static, spherically symmetric, perfect fluid solutions of Einstein's equations. Comp. Phys. Comm. **115**, 395 (1998)
25. G. Baym, H.A. Bethe, C. Pethick, Neutron star matter. Nucl. Phys. A **175**, 225 (1971)
26. H.A. Bethe, M. Johnson, Dense baryon matter calculations with realistic potentials. Nucl. Phys. A **230**, 1 (1974)
27. F. Weber, *Pulsars as Laboratories for Nuclear and Particle Physics* (IoP Publishing, Bristol, UK, 1999)
28. N.U. Bastian et al., Towards a unified quark-hadron-matter equation of state for applications in Astrophysics and heavy-ion collisions. Universe **4**, 67 (2018)
29. C.E. Rhoads, R. Ruffini, Maximum mass of a neutron star. Phys. Rev. Lett. **32**, 324 (1974)
30. A.Y. Potekhin, W.C.G. Ho, G. Chabriers, Atmospheres and radiating surfaces of neutron stars with strong magnetic fields. Proceedings of Science (MPCS2015). Preprint at arXiv:1605.01281
31. P.B. Demorest et al., A two-solar-mass neutron star measured using Shapiro delay. Nature **467**, 1081 (2010)
32. arXiv:2011.08157
33. R. Valentim, E. Rangel, J.E. Horvath, On the mass distribution of neutron stars. MNRAS **414**, 1427 (2011)
34. M.C. Miller et al., PSR J0030+0451 mass and radius from NICER data and implications for the properties of neutron star matter. Astrophys. J. Lett. **887**, L24 (2019)
35. A. Hewish et al., Observation of a rapidly pulsating radio source. Nature **217**, 709 (1968)
36. T. Gold, Rotating neutron stars as the origin of the pulsating radio sources. Nature **218**, 731 (1968)
37. F. Pacini, Rotating neutron stars, pulsars and supernova remnants. Nature **219**, 145 (1968)

38. F.C. Michel, *Theory of Neutron Star Magnetospheres* (University Chicago Press, Chicago, 1990)
39. P.M. Woods, C. Thompson, *Compact Stellar X-ray Sources* (Cambridge University Press, Cambridge, UK, 2006)
40. B. Davies et al., The progenitor mass of the magnetar SGR1900+14. Astrophys. J. **707**, 844 (2009)
41. G.S. Bisnovatyi-Kogan, B.V. Komberg, Pulsars and close binary systems. Sov. Astron. **18**, 217 (1974)
42. K. Hurley et al., An exceptionally bright flare from SGR 1806–20 and the origins of short-duration gamma-ray bursts. Nature **434**, 1098 (2005)
43. J.E. Grindlay et al., High-resolution X-ray imaging of a globular cluster core: Compact binaries in 47Tuc. Science **292**, 2290 (2001)
44. C. Montgomery, W. Orchiston, I. Whittingham, Michell, Laplace and the origin of the black hole concept. Jour. Astron. Hist. Heritage **12**, 90 (2009)
45. V.M. Kaspi, Grand unification of neutron stars. PNAS **107**(16), 7147–7152 (2010)
46. V.M. Kaspi, *Diversity in Neutron Stars: X-Ray Observations of High-Magnetic-Field Radio Pulsars* (American Astronomical Society, 2011)
47. B. Carter, Half century of black-hole theory: from physicists' purgatory to mathematicians' paradise, in *AIP Conference Series, Vol. 841*, Eds L. Mornas and J. Diaz Alonso, p. 29. arXiv:gr-qc/0604064
48. S.W. Hawking, Black hole explosions? Nature **248**, 30 (1974)
49. P.S. Custódio, J.E. Horvath, Evolution of a primordial black hole population. Phys. Rev. D **58**, 023504 (1998)
50. B. Carr, F. Kuhnel, M. Sandstad, Primordial black holes as dark matter. Phys. Rev. D **94**, 083504 (2016)
51. P.S. Custódio, J.E. Horvath, The evolution of primordial black hole masses in the radiation-dominated era. Gen. Rel. Grav. **34**, 1895 (2002)
52. https://eventhorizontelescope.org/
53. The Event Horizon Telescope Collaboration, First M87 Event Horizon Telescope results. I. The shadow of the supermassive black hole. Astrophys. J. Lett. **875**, L1 (2019)
54. https://stellarcollapse.org/bhmasses
55. J.A. Orosz et al., A 15.65-solar-mass black hole in an eclipsing binary in the nearby spiral galaxy M 33. Nature **449**, 872 (2007)
56. R. Narayan, J. Heyl, On the lack of type I X-ray bursts in black hole X-ray binaries: Evidence for the event horizon? Astrophys. J. Lett. **574**, L139 (2002)
57. R. Narayan, Black holes in Astrophysics. New J. Phys. **7**, 199 (2005). https://doi.org/10.1088/1367-2630/7/1/199
58. A.M. Ghez et al., Measuring distance and properties of the Milky Way's central supermassive black hole with stellar orbits. Astrophys. J. **689**, 1044 (2008)
59. Y. Tanaka et al., Gravitationally redshifted emission implying an accretion disk and massive black hole in the active galaxy MCG-6-30-15. Nature **375**, 659 (1995)
60. I.F. Mirabel, L.F. Rodríguez, A superluminal source in the Galaxy. Nature **371**, 46 (1994)

Chapter 7
Accretion in Astrophysics

7.1 Roche's Problem

The study of the accretion phenomenon started, in fact, as an orbital dynamics problem. In 1873 the French astronomer E. Roche formulated this concept for the first time, linking it to the form of the interior equipotentials in a binary system. Roche was interested in the fate of point satellites in orbit, but the idea is more general and can be applied to a particle distribution, for example, in a gas, as we will see below. The features of accreting systems, so important in Astrophysics, are presented and described.

The problem of perturbations on a body (star or planet) by the gravitational effect of a companion had already been considered before, but in the limit where these deformations were small. A clear example is the problem of induced tides. Deformations by rotation also fall into this category. In both cases the deformations are small and can be treated by proposing a series in axisymmetric Legendre polynomials. But Roche's problem does not belong to this category because the induced deformations are very big and need their own treatment [1].

To begin with, we establish the treatment in the context of the two-body problem (Fig. 7.1). The x-axis is placed along the line joining the two masses, and we assume that:

1. the system is in synchronous rotation,
2. the orbit is circular,
3. the masses are pointlike.

We consider a third body whose dynamics we intend to study, with position (x, y, z) and distances r_1 and r_2 relative to each of the other two masses given by

$$r_1^2 = x^2 + y^2 + z^2 , \quad r_2^2 = (x - a)^2 + y^2 + z^2 .$$

© The Author(s), under exclusive license to Springer Nature Switzerland AG 2022
J. E. Horvath, *High-Energy Astrophysics*, Undergraduate Lecture Notes in Physics,
https://doi.org/10.1007/978-3-030-92159-0_7

Fig. 7.1 Two-body problem
with masses M_1 and M_2 on a
synchronous orbit

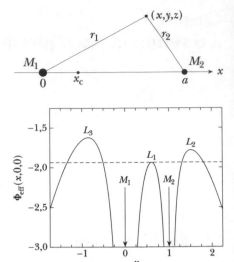

Fig. 7.2 A cross-section of
the potential (7.2) on the
x-axis. The situation is
similar to that of a small ball
on top of a hill. If it has
enough energy, it will fall
into the indicated "wells",
which have sizes
proportional to the masses

Using Kepler's third law, we can write $\omega^2 = (2\pi/P)^2 = GM/a^3$, with $M = M_1 + M_2$. Moreover, the center of mass of the system is located along the line that joins the two with x-coordinate x_c given by $x_c/a = M_2/M$.

We now define $q \equiv M_2/M_1 \leq 1$ and write $x_c/a = q/(1+q)$. In the rotating system the effective potential felt by the test particle will then be

$$\Phi_{\text{eff}}(x, y, z) = -\frac{GM_1}{r_1} - \frac{GM_2}{r_2} - \frac{\omega^2}{2}\left[(x - x_c)^2 + y^2\right], \qquad (7.1)$$

where the last term is the "centrifugal" term induced by rotation. Thus, using the relationships arising from Kepler's law we can make the potential dimensionless by replacing $x \to x/a$ to give

$$\Phi_{\text{eff}}(x, y, z) = -\frac{2}{(1+q)r_1} - \frac{2}{(1+q)r_2} - \left(x - \frac{q}{1+q}\right)^2 + y^2, \qquad (7.2)$$

which is *independent* of the individual masses and orbit size, since q is the only parameter that appears here. The solution of any problem involving (7.2) will be universal, because there is no reference to the dimensions in it. Once solved in terms of dimensionless quantities, it will be enough to restore the necessary dimensions for the specific problem at hand.

A cross-section of the potential is shown in Fig. 7.2. When placed in this potential, any particles with the energy to reach the "top" will inevitably fall to the other side. This means that when the deformation of the star is too big, matter (gas) will escape from it (M_2) and fall into the potential well of the other. Stellar Evolution will naturally place it in this secondary M_2 situation, and thus trigger the mass transfer.

A complementary perspective on this problem is given by calculating the points where a particle is subject to a balance of forces, i.e., the solutions of $\nabla\Phi_{\text{eff}} = 0$. In the (x, y) plane, the points that satisfy this equation are called *Lagrangian points*. The most important for our discussion is L_1, where the gravitational pull of the two masses is evident. But the other Lagrangian points are also relevant in practice. For example, the WMAP satellite mission and the *Herschel Observatory* were placed in orbits corresponding to point L_2 of the Sun–Earth system in order to maintain their orientation towards the Sun and to observe pointing in the opposite direction. Points L_4 and L_5 are a direct consequence of the rotation in the effective potential, without which there would be no forces to compensate gravitation. The so-called *Trojan asteroids* in the Sun–Jupiter system orbit at precisely these points.

In general, it is the joint existence of Lagrangian points and Stellar Evolution that leads to the phenomenon of accretion. We have already seen in Chap. 4 that solar type stars "swell" considerably when they leave the Main Sequence and need to conserve Virial equilibrium and energy simultaneously. Thus, the gas in the atmosphere of the star at some point reaches L_1 and one says that the secondary "fills its Roche lobe". This is the moment where accretion begins in the binary system.

Of course, this is only possible if the deformation of the secondary star is very large. In fact, the atmosphere fills an equipotential like those marked in light blue in Fig. 7.3. The exact calculation of the shape of these equipotentials is difficult, but there is a numerical fit due to Eggleton [2] for the quotient of the Roche lobe R_L and the semi-axis a as a function of the mass asymmetry q which turns out to be very precise in any situation and is used for a variety of calculations:

$$\frac{R_L}{a} \approx \frac{0.49\,a^{2/3}}{0.69\,a^{2/3} + \ln(1 + q^{1/3})} \,. \tag{7.3}$$

Numerically, the equipotentials can be calculated without difficulty using this expression as a function of the mass asymmetry q. Figure 7.4 shows a secondary filling the Roche lobe for $q = 1$. It is clear that the initial discussion about the magnitude of the deformations was fully justified.

In general, if the primary is a compact object it will never be able to fill its Roche lobe. But in cases where the binary also features at least one "normal" star (i.e.,

Fig. 7.3 Roche potential in the (x, y) plane. The masses M_1, M_2 with $q = 3$ are in the dark regions, and the five Lagrangian points are indicated

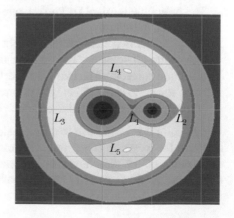

Fig. 7.4 Roche lobe of the secondary (*red*) filled in a binary with $q = 1$ and small inclination with respect to the line of sight. Note the considerable deformation pointed out at the beginning of the chapter

Fig. 7.5 The four types of binary depending on the filling of the Roche lobe [1]. *Top left*: Detached. *Top right*: Semi-detached. *Bottom left*: Contact binary. *Bottom right*: Common envelope. Credit: Phillip D. Hall CC BY-SA 4.0

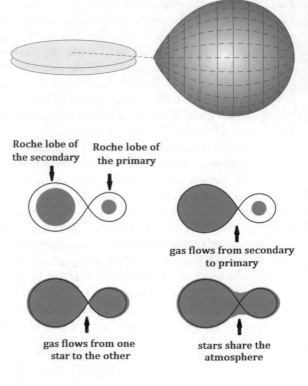

non-degenerate star), there are several possibilities for filling. Thus, binaries can be classified as follows by the way they fill their respective Roche lobes (Fig. 7.5):

1. Separated (detached). Neither member fills its Roche lobe. This type of binary is frequent for pairs with a long period, but can be converted to another type later depending on the evolution of the members.
2. Semi-detached. The secondary fills its Roche lobe and there is transfer of mass from it to the primary. Typical case of a binary with a compact object as primary. The prototype is Algol (β Persei), one of the first variable stars identified in the history of Astronomy.
3. Contact. The two stars fill their Roche lobes and the gas from one or the other flows through point L_1. The prototypical example is W Ursa Majoris.
4. Common envelope. The two stars overflow their respective Roche lobes and thus share the atmosphere. This type of binary has a very short duration, a few years at most, before expelling the envelope or merging.

At this point it is instructive to take a look at Fig. 7.13, where the secondary Roche lobes in the systems containing a black hole are shown to scale. The physical dimensions of the disks are also important because they determine the existence of *eclipses*, as we discussed in the case of M33 X7.

7.2 Spherical Mass Accretion and Accretion Disks

In general, the fall of matter on the primary leads to one of the most common phenomena in Astrophysics: the so-called *accretion disks*. It is not so obvious that matter should form a disk. In fact, it could fall in an isotropic (spherical) manner. But the main factor determining each possibility is the *angular momentum* of the gas in the potential: spherically symmetrical accretion requires the gas to have zero total angular momentum, which is very unlikely to happen. With this caveat, we first present the elementary treatment of spherical accretion (zero total angular momentum) before discussing the disks.

In the simplest form of spherical accretion, called Bondi–Hoyle accretion, an object of mass M travels through the interstellar medium with a velocity v, in a subsonic ($v < c_s$) or supersonic ($v > c_s$) way. The isotropic fall of particles onto the object causes an addition of mass that obeys the simple expression

$$\dot{M} \approx \pi R^2 \rho v \,, \tag{7.4}$$

where the right-hand side is just the spherical flux multiplied by the geometrical cross-section πR^2. Equation (7.4) uses the speed of the object v if the motion is supersonic, while the speed of sound c_s must be inserted in place of v if it is subsonic. Note that the radius R in (7.4) is not exactly the radius of the object, but rather an effective radius (the *Bondi radius*), determined by the escape condition for the speed of sound c_s, given by $\sqrt{(2GM/R)} = c_s$. This expression thus defines the Bondi radius by $R_B = 2GM/c_s^2$, and this can be substituted into (7.4) to obtain

$$\dot{M} \approx \frac{\pi \rho G^2 M^2}{c_s^3} \,. \tag{7.5}$$

This formula shows that the higher the speed of the object, the less mass will be accreted, besides depending quadratically on the central mass. The full problem is much more complicated and needs a detailed treatment in the neighborhood of the object where the accreted gas suffers from discontinuities in density that appear because the supersonic flow has to "fit" the boundary conditions, but we will not address these complications in our discussion.

The more general case of non-spherical accretion begins with the observation that, when the angular momentum is not zero, its (approximate) conservation will cause a *flattening* of the gas flow near the attracting mass M. As the gas must be able to cool down faster than it loses angular momentum (this is a precondition for it to fall), a flattened rotating structure is formed which we call a *disk* [3].

The example shown in Fig. 7.6 is the disk of the object Herbig–Haro 30. These systems correspond to stars that are just forming and where the central core is accreting matter from the cloud that builds the star. The disk is in the equatorial plane, perpendicular to the proto-stellar jets, as expected.

Fig. 7.6 HH 30, a star in an early stage of formation. The disk occupies the equatorial plane and scatters photons into its neighborhood. The jets result from the same angular momentum transport Physics that causes the formation of a disk. Credit: C. Burrows (STScI and ESA), the WFPC 2 Investigation Definition Team and NASA

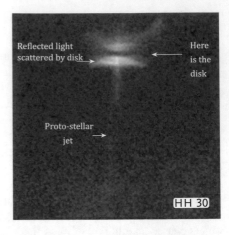

Reflected light scattered by disk

Here is the disk

Proto-stellar jet

HH 30

Being a little more specific, we can study when the disk will form in terms of the angular momentum J, equating the gravitational force to the centrifugal force resulting from it:

$$\frac{GMm}{R^2} = m\omega^2 R . \tag{7.6}$$

Defining the *circularization radius* $R_{\mathrm{circ}} = J'^2/GM$, where $J' \equiv J/m = \omega^2 R$ is the angular momentum per unit mass, we find that the disk must form when R_{circ} is greater than the radius R of the primary. If this condition is not fulfilled, the disk does not form. But this is precisely what happens in the overflow of the Roche lobe discussed above: the angular momentum of the gas is too great for it to reach the central object directly. Therefore, the disk is a corollary of this excess angular momentum.

The next issue is the structure of this disk. Let us imagine the disk divided into concentric rings (Fig. 7.7 left). Because of its proximity to the center, the ring A rotates faster than the ring B. Thus, the existing friction between the two tends to brake A and accelerate B. This friction can be dissipative (since pure rotation could not make matter "fall" towards the central object) or, if not effective, the angular momentum must be transported radially for the accretion to occur. A disk with pure Keplerian rotation of the gas particles without interaction (that is, where $GM/R^3 = \omega^2$) has stationary orbits, and accretion never occurs. Therefore, the angular momentum *cannot* be strictly conserved; it must be dissipated (radiated) or transported. Figure 7.7 (right) shows the rotation profile $\omega(R)$ of the gas of a Keplerian disk. Of course, matter rotates faster near the center, going as $R^{-3/2}$, only non-ideal effects can cause its effective fall, and the disk is precisely the place where this happens.

To model the effect of dissipation, we introduce a *viscosity* in the disk gas. The physical origin of this viscosity presents a serious problem, since the simplest estimates show that it must be orders of magnitude greater than the "natural" *ansatz* value of the molecular viscosity, with kinetic expression $\nu_{\mathrm{mol}} = \lambda_{\mathrm{mol}} \times c_{\mathrm{s}}$. This

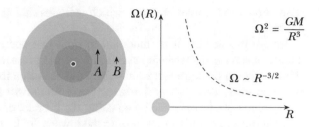

Fig. 7.7 *Left*: Scheme of rings that form an accretion disk. *Right*: Matter rotation profile of a Keplerian disk

conclusion follows from the fact that, as $\lambda_{mol} \approx 10^{-2}$ cm and the speed of sound in the medium is $c_s \approx 10^6$ cm s^{-1}, the characteristic time for changes in the disk should be

$$\tau_{mol} = \frac{R^2}{\nu_{mol}} \sim 10^8 \text{ yr} . \tag{7.7}$$

On the other hand, observations of real systems with accretion disks show important variations of the order of weeks (!). Thus, it is very likely that there are viscosity sources orders of magnitude higher acting in the disk, in such a way as to produce these variations. At the moment, the exact physical origin of this viscosity is quite unknown.

To solve this problem and construct a model of the disk, Shakura and Syunyaev [4] suggested dealing with the low value of the molecular viscosity as follows: a low viscosity ν_{mol} will necessarily cause the regime change from laminar to turbulent gas flow. The reason for this is that the *Reynolds number* $Re = vL/\nu_{mol} \approx 10^{10}$ is huge, much higher than the threshold value to consider the gas flow as laminar. Thus, they deduce that the main effect on the disk would be due to *turbulent viscosity* ν_{turb}. However, the theory of turbulence is complicated and the viscosity cannot be calculated from first principles. Thus, Shakura and Syunyaev [4] parametrize the turbulent viscosity as $\nu_{turb} = \alpha c_s H$, where H is a height scale for the pressure in the disk, i.e., the distance at which the pressure in the transverse direction drops appreciably. Using the hydrostatic equilibrium equations, continuity equation, energy balance, etc., giving 6 differential and algebraic equations in all, the disk problem can be solved with $\alpha \sim 1$ as a parameter. The physically important hypotheses for these so-called α-disks are:

- The gravitational field is due to the central object alone, i.e., the disk is not "self-gravitating".
- The disk is geometrically thin, but optically thick.
- Hydrostatic balance determines the vertical structure, and in particular, the height scale H.
- There are no external winds or torques on the axisymmetric disk.

Such α-disks are widely used in the absence of a more precise model for the sources of turbulence and other effects. In fact, the initial proposal was replaced by the hypothesis that the viscosity is proportional to the gas pressure, since if it were proportional to the total pressure, which includes radiation, the situation would be unstable.

Before proceeding, it is important to note that accretion cannot be arbitrarily intense, since the matter being accreted feels the pressure of the emitted radiation. If the radiation pressure that results from the accretion itself becomes too high, the accretion stops. There is therefore a *maximum luminosity* for any object that accretes mass, which can be obtained by equating the product of the radiation pressure and the (Thompson) cross-section σ_T with the inward "pull" of the gravitation, i.e.,

$$\frac{\mathrm{d}p}{\mathrm{d}t} = \sigma_T \times \frac{L}{4\pi c r^2} \ . \tag{7.8}$$

where $\mathrm{d}p/\mathrm{d}t = -GMm_p/r^2$, with m_p the mass of the proton (hydrogen nucleus). Solving for the luminosity that satisfies this equality, we obtain the maximum luminosity as

$$L_E = \frac{4\pi G M m_p}{\sigma_T} \ , \tag{7.9}$$

known as the *Eddington luminosity*, the maximum possible value allowing the disk to remain bound. As a corollary of this idea, we see that every explosive phenomenon (for example, X-bursts, etc.) must be *super-Eddington* [5].

As listed above, the second hypothesis put forward by Shakura and Syunyaev, that the disk should be optically thick, implies that it will emit as a whole as a black body, with peak temperature given by $L_{\mathrm{disk}} = 4\pi R^2 T_{\mathrm{peak}}^4$. If we suppose that this temperature corresponds to the matter emitting in the last stable orbit with $R = 3R_S$ (Chap. 6), and that the maximum luminosity is the Eddington luminosity L_E, we can invert to find T_{peak} as a function of the mass of the central object:

$$T_{\mathrm{peak}} = 2 \times 10^7 \left(\frac{M}{M_\odot}\right)^{-1/4} \mathrm{K} \ . \tag{7.10}$$

Thus, we see that the highest temperatures are reached for the less massive objects. For example, a supermassive black hole disk will emit in the UV, but a microquasar disk will do so mainly in X-rays. Note that this statement applies only to the emission of the disk, while other components such as a corona and jets may be present and add to the total radiation emitted in other bands.

With the advances in the study of the fundamental Physics of accretion it became clear that the viscosity can be attributed to the presence of magnetic fields. Magnetized disks are subject to magnetohydrodynamic (MHD) instabilities that may be responsible for the observed *changes in accretion state*, like those shown in Fig. 7.8.

One important feature that is known to be present in compact objects such as neutron stars is a high magnetic field. The magnetic field of the central object is expected to affect the accretion itself, because at some point the magnetic pressure would oppose the further fall of matter. Assuming a dipolar structure for the field $B \propto (R/r)^3 B_0$, where B_0 is the intensity at the surface $r = R$, this magnetic pressure

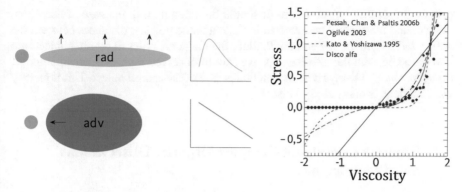

Fig. 7.8 *Left*: Changes in the accretion disk observed on weekly time scales. The disk goes from emitting as a black body to a state where the emission is a power law, with alteration to its structure. *Right*: Calculated curves due to Pessah, Chan, and Psaltis, taking into account turbulent MHD instabilities, compared with the Shakura–Syunyaev parametrization (*red line*). These differences are surely what lies behind the sudden changes observed. Credit: Fig. 1 of [6]

reads

$$P_{\text{mag}} = \frac{B^2}{8\pi} = \left(\frac{R}{r}\right)^6 B_0^2 . \tag{7.11}$$

On the other hand, the ram pressure of freely falling matter is

$$P_{\text{ram}} = \rho v^2 = \frac{\dot{M}}{4\pi r^2}\left(\frac{2GM}{r}\right)^{1/2} . \tag{7.12}$$

At some point near the accreting object, the magnetic pressure will dominate and the matter will obey the dynamics imposed by the spatial structure of the magnetic field. This point is called the *Alfvén radius*, given by

$$r_{\text{A}} = \left(\frac{8\pi^2}{G}\right)^{1/7}\left(\frac{R^{12}B_0^4}{M\dot{M}^2}\right)^{1/7} . \tag{7.13}$$

Since the central object is generally rotating, there is another relevant quantity for the accretion onto a magnetized compact star, namely the *corotation radius*, denoted by r_{co}, the locus of points at which the corotation is at the maximum possible speed c:

$$r_{\text{co}} = \left(\frac{GM}{\Omega}\right)^{1/3} . \tag{7.14}$$

This has to be compared to the Alfvén radius to gauge the effects of the magnetic field and rotation together: if $r_{\text{A}} < r_{\text{co}}$ the matter accretes onto the object, but if $r_{\text{A}} \gg r_{\text{co}}$, the matter will be ejected. This last situation is called a *propeller*, when the mass gain by the central object is zero. It should be noted that, in the case of accretion, the accreted matter will be funneled to the magnetic poles at distances between the Alfvén radius and the star surface R. Since the magnetic poles in rotating objects are never aligned with the rotation axis, we should observed a periodic modulation of (*bremsstrahlung*) X-rays with the rotation period of the central object. This is exactly what happens is many X-ray pulsars.

7.3 Binaries Containing Compact Objects: Observations and Classification

Our discussion above has been quite general and most of it can be applied to any accreting binary. However, as we have a special interest in binaries where at least one of the members (the primary) is a compact object, we can use the previous concepts to classify and understand observations. We will now carry out a brief overview of these systems.

7.3.1 Cataclysmic Variables (CV)

This class of variables containing a white dwarf has been known for several centuries, originally through the events known as *novas*, stars that greatly increase their usual magnitude and then return to their previous state. These events consist of outbursts in which the brightness of the system can increase by 10 magnitudes or more. It was later found that this class contains several subgroups consisting of a semi-detached binary where M_1 is a white dwarf.

The mechanism by which the nova produces an optical outburst is now well accepted: the accreted hydrogen accumulates on the surface and reaches an ignition condition, producing the event. Different systems are characterized by different secondaries, WD magnetization, etc. Note that, if the WD retains a fraction of the accreted mass, it may become an SNIa progenitor if it approaches the Chandrasekhar limit. This is in fact the proposed fate of V4444 Sgr, shown in Fig. 7.9.

The CV zoo contains several interesting species that can reveal details of the accretion process and its consequences, e.g., the outbursts, but we will not discuss them here. We recommend the article by R. Connon Smith [7] for an in-depth discussion.

Fig. 7.9 Typical outburst of the variable V4444 Sgr, with nova eruptions discovered in the 19th century. Note the increase in brightness by almost 7 optical magnitudes. Credit: American Association of Variable Star Observers (AAVSO)

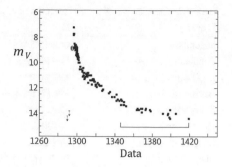

7.3.2 Low-Mass X-Ray Binaries (LMXB) and High-Mass X-Ray Binaries (HMXB)

A significant number of binary systems harbor neutron stars: we have already mentioned several relativistic binaries (WD–NS, NS–NS), but X-ray binaries can also have "normal" stars as donors. This is the case of the so-called *low-mass X-ray binaries* and *high-mass X-ray binaries* (LMXB and HMXB). In both cases the compact object can be either an NS or a BH, and it is sometimes difficult to discriminate which is which. The simplest estimate of luminosity in X-rays due to the accretion of gas from the companion does not depend on the mass of the donor, but on the matter transfer rate and on the mass M_X of the compact object, according to

$$L_X = \frac{GM\dot{M}_X}{R_X} ,$$ (7.15)

i.e., the kinetic energy of the accreted gas is converted into radiation and heat on or near the surface of the compact object. We are not considering the possibility of accretion disk states that are optically thin, where the estimate (7.15) is not valid. Note that the accumulation of gas, a necessary condition for thermonuclear outbursts, is possible only if there is a surface, so black holes could not have outbursts (Chap. 6). There are several striking differences between LMXB and HMXB, as noted in Table 7.1 [5].

On general grounds we can say that the nature of the secondary star mainly determines the type of binary, while the intense magnetic field of the neutron star causes the gas to fall essentially through the poles, directing the flow (and justifying the presence of X-ray pulsations resulting from stellar rotation); the weak field of the LMXB does not favor pulsations (Fig. 7.10), but the accretion flow presents a series of quasi-periodic oscillations (QPO) in the spectrum that reveal important signatures of the inner regions of the disk, and possibly of the compact star itself, a topic of great interest to the community.

An extreme example of systems that started as a type of LMXB but have reached a very advanced stage of evolution are the so-called *black widows* (Fig. 7.11). The

Table 7.1 Comparison between HMXB and LMXB

Property	HMXB	LMXB
Secondary	O-B ($>5M_\odot$	K, M, or WD
Population	I ($\sim 10^7$ yr)	II ($\sim 10^9$ yr)
L_X/L_{opt}	10^{-3}–10	100–1000
Accretion disk	Yes (?) transient	Yes
X-ray spectra	Hard ($k_B T > 15$ keV)	Soft ($k_B T \sim 10$ keV)
Orbital period	1–100 days	10 min–10 days
Eclipses (X-rays)	Common	Rare
Magnetic field	Strong ($\sim 10^{12}$ G)	Weak ($\sim 10^7$–10^8 G)
Pulses	Common (0.1–1000 s)	Not so common (0.1–100 s)

Fig. 7.10 A typical LMXB, where the central object can be a neutron star or a black hole. The fundamental elements of the binary system are indicated

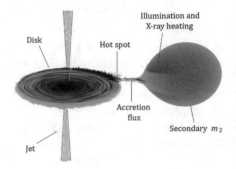

discovery of these systems [8] immediately led to the construction of a model where the observed eclipses were associated with the *evaporation* of the companion by the pulsar wind. This wind-stripped matter is easily visible in images, giving the evaporating star the aspect of a comet—and in fact, this is exactly what happens with cometary nuclei when they approach the Sun, only on a much higher energy scale. The name "black widow" signals the fact that it is the very star that is being evaporated that was responsible for re-energizing the pulsar in a previous phase. The current mass of the companion in this and similar systems is 10^{-2}–$10^{-3} M_\odot$, and will possibly be completely evaporated in the future. Years after the original discovery, an Australian group found similar binary systems, but where the mass of the companion was of order $0.1 M_\odot$, and named them *redbacks* (an Australian spider related to the black widow species). The evolutionary explanation for these systems appeals to X-ray backlighting in the redback phase and the evolution of the secondary until it later reaches the condition of degeneracy: the degenerate matter *expands* if the mass is ablated by the wind, so the object is subject to further evaporation. This makes some redbacks the progenitors of black widow systems, with a unique evolution combining backlighting and wind in systems that reach orbits with periods of a few hours [9]. These groups are an area of intense research to understand and confirm their structure and evolution.

Fig. 7.11 Composite image of the original black widow system (B 1957+20). The X-ray emission in *red* is the ablated matter stripped from the surface of the companion by the pulsar wind, located very close to the *white point*. On this scale, it is impossible to resolve it in the images (the orbital period is only 9.2 hours). Credit: X-ray NASA/CXC/ASTRON/ B. Stappers et al., optical AAO/J. Bland-Hawthorn & H. Jones

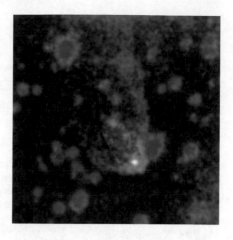

The denomination *high-mass X-ray binaries* (HMXB) applies to a class of systems in which a massive star, sometimes even a Wolf–Rayet type, is present as a donor. The known existence of winds from these objects in the advanced stages creates many complications for their study. Some of them are related to heavy obscuration and, even if multi-spectroscopy can be performed, the status of their atmospheres is quite complex and requires very detailed models. On the whole, it is not uncommon to find highly variable X-ray emission, sometimes attributed to the decay and reconstruction of a disk in systems of high eccentricity. If anything, the HMXB systems are more difficult to model and understand than their "cousins" the LMXB [10].

7.3.3   More on the Binary Systems Containing Black Holes

The identification of black holes in LMXB/HMXB deserves some additional remarks. Until 1990 or so, the consensus regarding the presence of black holes in binaries was determined by the shape of the spectrum. The "hard" spectrum was related to black holes, while the "softer" form was associated with neutron stars. But when more accurate data became available, it was found that there can be *transitions* between two such states in the same source. Not only do binary black holes suffer this transition, but so do others where a neutron star is the central object. Thus, a definitive identification by means of spectral data alone is not really possible (Fig. 7.12).

In the "soft" state, it is believed that the X photons are produced and escape from the innermost part of a very fine accretion disk, with a quasi-thermal spectrum, while in the "hard" state, the photons are reprocessed by a more spherical "corona" and the spectrum is of the power-law type, and there is radio emission, indicating the presence of jets. Changes in one direction or the other have been observed between these two states, and thus associated with changes in the state of the disk, possibly due to instabilities in the latter induced by the radiation that illuminates it. We see

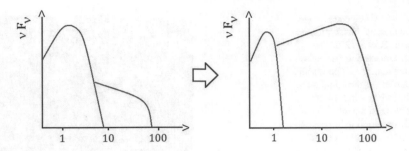

Fig. 7.12 The kind of spectral changes seen in some X-ray sources, e.g., GRO J1655-40. *Left*: A so-called soft-high state dominated by thermal emission from the disk. *Right*: A hard-low state with a non-thermal component extending typically above 100 keV. These changes go back and forth with frequency in the same source and are thought to reflect changes in the disk

Fig. 7.13 X-ray emitting binary systems that contain black holes. Note the scale of Mercury's orbit at the top. For some of these systems, the "year" lasts only a few hours. The disks correspond to their actual size properly scaled [12]. Credit: Jerome Orosz, San Diego State University

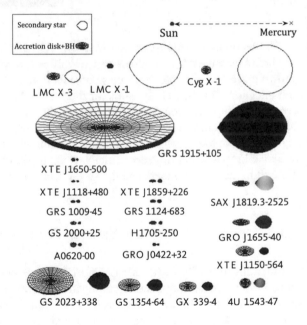

here how important it is to know the state of the disk, both for its own sake and when we use it to diagnose the central object.

Figure 7.13 shows to scale the binary systems in which a black hole is considered to be present [12], usually because the estimated mass from Kepler's third law exceeds the Rhoades–Ruffini limit given in (6.37). Due to the different parameters of binary formation and the various evolutionary states of the companions, black holes that emit X-rays form a particularly interesting set when we look at them on the same scale as in Fig. 7.13, where the various systems have been classified according to the color of the companion star (obtained from the effective temperature), the size of the measured orbit, and the observed scale of the accretion disk.

Much remains to be done regarding these systems, whose estimated number in the Galaxy could reach 10–100 million. X-ray instrumentation and ongoing spectroscopic and photometric surveys will be very important in this regard.

References

1. B.W. Carroll, D. Ostlie, *An Introduction to Modern Astrophysics* (Cambridge University Press, Cambridge, UK, 2017)
2. P.P. Eggleton, Aproximations to the radii of Roche lobes. Astrophys. J. **268**, 368 (1983)
3. M. Abramowicz, Introduction to the theory of black hole accretion discs. AIP Conf. Proc. **431**, 34 (1998)
4. N.I. Shakura, R. Syunyaev, Black holes in binary systems: Observational appearance. Astron. Astrophys. **24**, 337 (1973)
5. S. Shapiro, S.L. Teukolsky, *Black Holes, White Dwarfs and Neutron Stars: The Physics of Compact Objects* (Wiley-VCH, Weinheim, 1991)
6. M.E. Pessah, C.-K. Chan, D. Psaltis, MNRAS **383**, 683 (2008)
7. R. Connon Smith, Cataclysmic variables. Contemp. Phys. **47**, 363 (2006)
8. A.S. Fruchter, D.R. Stinebring, J.H. Taylor, A millisecond pulsar in an eclipsing binary. Nature **333**, 237 (1988)
9. O.G. Benvenuto, M.A. De Vito, J.E. Horvath, Understanding the evolution of close binary systems with radio pulsars. Astrophys. J. Lett. **786**, L7 (2014)
10. A.A.C. Sander, Massive star winds and HMXB donors, in High-mass X-ray binaries: Illuminating the passage from massive binaries to merging compact objects. Proc. Int. Astron. Union **346**, 17 (2019)
11. R.A. Remillard, J.E. McClintock, Active X-ray states of black hole binaries: Current overview. Bull. Am. Astron. Soc. **38**, 903 (2006)
12. J. Orosz, http://chandra.harvard.edu/graphics/gallaryxray.gif

Chapter 8
Active Galactic Nuclei (AGNs)

8.1 Discovery of Quasars

As we saw in Chap. 2, the launch of the first satellites carrying X-ray detectors into space in the 1960s was a game-changer in high energy Astrophysics. Added to the prodigious development of radio telescopes, a flood of completely unexpected discoveries revealed the existence of large amounts of energy emitted by objects with a stellar appearance, but whose extragalactic origin was eventually established. These were thus called quasi-stellar objects (QSO) or *quasars*. The first of these objects, the most famous example of a quasar, called 3C 48, was discovered in the 1960s. Initially, it could not be classified on the basis of its observed spectrum. A general discussion of the AGN types and evolution is given here. A series of rather broad lines in positions that did not correspond to known elements left astronomers puzzled. Soon after, similar objects were found. In fact, the original name "quasar" already shows how difficult it was to understand these objects, since they look like any ordinary star to an optical telescope.

However, an in-depth study of their spectral lines, produced by atomic absorption and emission, which removes or adds photons in certain regions of the wavelength range, showed that these lines were actually due to known transitions of ordinary chemical elements. It was just that they were located in very different positions from the usual ones, being *uniformly* shifted toward the red region of the spectrum (Fig. 8.1). This uniform displacement was soon attributed to the *expansion of the Universe*, which in turn implied that quasars were located at cosmological distances, with the emitter gas corresponding to some type of galaxy. The work that allowed this conclusion had begun 50 years earlier when V. Slipher measured the radial velocities of several "nebulae" [1]. It was later shown that the "nebulae" were actually giant stellar systems (galaxies), and that the expansion of the Universe would cause what appeared to be a "Doppler effect" between them and the Earth due to their relative velocity v. This was thus observed as a uniform displacement of the lines from the laboratory position λ_0 to the observed position λ according to [2]

© The Author(s), under exclusive license to Springer Nature Switzerland AG 2022
J. E. Horvath, *High-Energy Astrophysics*, Undergraduate Lecture Notes in Physics,
https://doi.org/10.1007/978-3-030-92159-0_8

Fig. 8.1 The redshift of the emission lines of four quasars, progressively more distant *from top to bottom*. Note the uniformity of the displacement that affects all lines equally. Credit: C. Pilachowski, M. Corbin/ NOIRLab/NSF/AURA

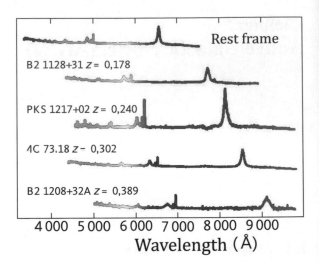

$$z = \frac{\lambda - \lambda_0}{\lambda_0} = \frac{v}{c}. \tag{8.1}$$

A little later, the Friedmann–Robertson–Walker cosmological models were developed. These contain solutions in which the scale of the Universe increases with time, and it became clear that, in an expanding cosmological context, the spectral shifts were *not* actually due to the Doppler effect, but rather to the very expansion of space-time that "drags" the quasar with respect to Earth. That is, it is not that there is a relative velocity v while the space-time remains fixed, but it is the space-time itself that expands (increases in scale), taking with it all the objects in the Cosmos. However, to first order in v/c, the expression (8.1) is correct because it is the first term in a Taylor series, and can be used at least for small displacements. The idea of a "Doppler effect" has stuck, largely due to the interpretation by De Sitter and others, but it is conceptually wrong according to our present understanding.

After a while astronomers became quite convinced that quasars were extragalactic objects, although this meant that the emitted energy had to be extremely high. The rapid variations observed in the light curves (on a short time scale $\tau < $ minutes) showed that they were very compact (Chap. 3), because as discussed previously, the emission region had to be smaller than $c\tau$ to account for the observed variation without violating causality. But the problem of the energy source remained unsolved. One very popular idea was that these systems were fueled by annihilating matter and antimatter, but this required credible scenarios for the production of antimatter. At least two decades research were needed to conclude that the most viable model was in fact the accretion of matter by very massive black holes, with gigantic masses, up to 10^6 times the mass of the Sun. Only such high masses could explain the emitted energy (see below). Thus, QSOs were associated with another class of active galaxies, well known for emitting radio and X-rays in what is now known as the unified AGN model.

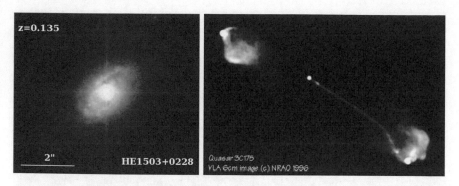

Fig. 8.2 *Left*: The quasar HE1239-2426 in the center of the host galaxy, revealed in high-resolution images. *Right*: Example of extragalactic jets in the quasar 3C 175. Credit: J. Bahcall (IAS, Princeton), M. Disney (Univ. Wales), NASA

The question that initially motivated these studies was: what makes QSOs visible? In other words, what is their energy source? After a long scientific controversy, working through all possible hypotheses, including those that made black holes responsible for the energy emission, the cosmological origin of quasars was finally established when a supernova was observed in one of them. Thus, it was proven that each quasar resided in some kind of galaxy in which there were also normal stars such as the progenitor of the supernova that exploded. Today we can see some of these galaxies in deep images like those in Fig. 8.2. Naturally, the stars in these galaxies were not visible individually because of the enormous distances involved. Relativistic particle jets that emerge from these regions and propagate through many kpc were also observed in AGNs as an additional product of mass accretion onto the black hole (Fig. 8.2). They are formed by matter that escapes perpendicular to the plane of the accretion disk in the neighborhoods of the central black hole, as discussed in the last Chapter. Everything pointed to the presence of the largest accreting systems in the entire Universe. It is the accretion of gas and stars by supermassive black holes that makes us see a quasar, while most of the time the host galaxy is invisible because it is very faint at typical distances [3].

To establish that the quasar energy source is *not* the annihilation of matter with antimatter as was initially thought, but rather the accretion of gas onto a gigantic black hole, it was important to estimate the mass of the black hole. Let us see how this is done in the case of AGNs. The luminosity produced by matter falling at a rate $\dot{M}$ is

$$L = \frac{GM\dot{M}}{2r} = 2\pi r^2 \sigma T^4 , \qquad (8.2)$$

where the factor of two arises from the conversion of energy according to the Virial theorem (4.27), energy which is then radiated by both faces of the disk behaving as a black body. Equation (8.2) can be inverted to obtain

Fig. 8.3 The galaxy Cygnus A in the optical band (*left*) and X-ray/radio (*right*) bands. Although to conventional telescopes there seems to be nothing special about it, when investigated in other bands both the X-ray emission (*blue*) and the emerging jets (*red*) place it firmly in the active galaxy class. Credits: X-ray NASA/CXC/SAO; visible light NASA/STScI; radio NSF/NRAO/AUI/VLA

$$T \propto M \dot{M}^{1/4} r^{-3/4} \,, \tag{8.3}$$

or, by introducing the Schwarzschild radius R_S from Chap. 6 as a suitable spatial scale when dealing with a black hole:

$$T = \left(\frac{3GM\dot{M}}{8\pi R_S^3} \right)^{1/4} \left(\frac{r}{R_S} \right)^{-3/4} \,. \tag{8.4}$$

Recalling that matter follows spiral paths until it suddenly falls when it reaches the radius of the innermost stable circular orbit (ISCO), which is $3R_S$ if we do not take into account the spin of the black hole, we come to the conclusion that the hardest radiation comes from the inner edge of the disk, while the optical and infrared signals are produced in the regions farthest out, where the temperature has dropped sufficiently. This applies to the disk radiation, but we have already seen that the radio emission is almost certainly due to the synchrotron provided by the jet electrons (although it is not totally clear where the magnetic field **B** that makes it possible comes from). This radio emission is also closely related to the harder emission in gamma rays, since the very energetic electrons from the jets collide with photons and transfer their energy to them by the inverse Compton process discussed in Chap. 2. But note that the exact location of these "soft" photons is still a subject of debate.

In any case, to reproduce the observed luminosities and assuming that the emission occurs at Eddington's maximum rate, (8.2) with $\dot{M} = \dot{M}_{\text{Eddington}}$ indicates that the central objects must have a mass M of several millions or even billions of solar masses in some cases. This means that we are dealing with the black holes described as *supermassive*, already located in the Carter diagram Fig. 6.26, and totally unrelated to the Stellar Evolution process.

It is not always easy to identify a quasar with an active galaxy or AGN. There are several types of active galaxy that appear to be normal in optical imaging, just a galaxy like any other. But when examined in radio, X-rays, or gamma rays, some of these emissions can be much larger than one would see from a normal galaxy. That is, some kind of high-energy process is active in the galaxy, without necessarily appearing in the optical bands. One of the best-studied examples is shown in Fig. 8.3.

8.2 Types of AGN and the Unified Model

With the appearance and classification of several types of AGN exhibiting different emissions, the idea gradually emerged that they might actually have the same basic structure, but observed from different angles due to their random orientations relative to the line of sight. The construction of this unified model of AGNs took into account the existence of a supermassive black hole, together with matter in an accretion disk, and also abundant dust (detected) that would obscure the system in the equatorial plane. Gas clouds in the inner and outer region were postulated to explain the widths of the lines (like those in Fig. 8.1). Narrow lines should originate far from the central region, where the clouds move at low speeds, while broad lines should be produced in closer regions where the clouds move in short orbits and hence more quickly. Depending on the inclination of the object's "equatorial" plane with respect to the line of sight, we will see different aspects of the active galaxy. In this way, the variety of types found was unified by astronomers, and generically called AGNs, encompassing quasars and other "cousins" of a similar nature (such as Seyfert galaxies, blazars, etc.), but which would ultimately turn out to be the same type of object [4].

Figure 8.4 displays the various types of AGN that needed to be explained. The basic classification begins with the observation of radio emission, which can be weak/null (the AGN is said to be radio-quiet) or significant (the AGN is said to be radio-loud). The presence or absence of wide or narrow lines also divides the AGNs into categories I and II, with a category 0 indicating the absence of both types (although there may be absorption lines). The Unified Model thus has to contain regions where narrow lines and wide lines are produced, and it has to explain why they appear or don't appear in the various groups.

The physical structures that were postulated to explain each observed characteristic are visualised in Fig. 8.5. Basically, the black hole-accretion disk (fossil) system occupies the central region or "engine" of the AGN, but is surrounded by a torus of gas and dust that substantially obscures the system in the equatorial plane, and clouds of gas in the perpendicular direction, in addition to jets that are mainly responsible for the radio emission. (Note that the torus of gas and dust should not be confused with the accretion disk itself, which is almost pointlike on the scale at which the figure is drawn.) According to this Unified Model, the division between "loud" or "quiet" then depends on the jets and the angle between the jet axis and the line of sight. A representation of what is observed in each case for different angles of observation is shown in Fig. 8.5.

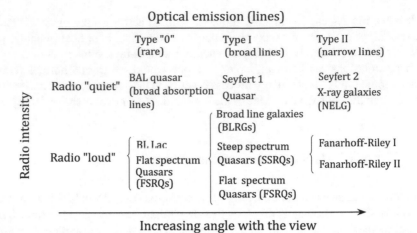

	Optical emission (lines)		
	Type "0" (rare)	Type I (broad lines)	Type II (narrow lines)
Radio "quiet"	BAL quasar (broad absorption lines)	Seyfert 1 Quasar	Seyfert 2 X-ray galaxies (NELG)
Radio "loud"	BL Lac Flat spectrum Quasars (FSRQs)	Broad line galaxies (BLRGs) Steep spectrum Quasars (SSRQs) Flat spectrum Quasars (FSRQs)	Fanarhoff-Riley I Fanarhoff-Riley II

Increasing angle with the view

Fig. 8.4 The AGN zoo

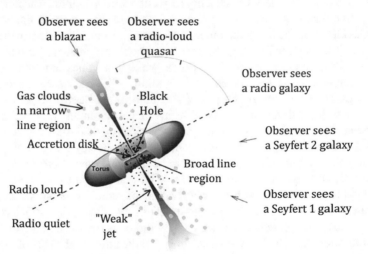

Fig. 8.5 Different structures featuring in the Unified Model of AGNs, and the result of observing them along different lines of sight [5, Fig. 1]

We should point out that, in addition to the often observed jets, quite direct evidence of the presence of the disk has been obtained, for example, by the instruments of the Hubble Space Telescope. The regions where the wide and narrow lines should form are inferred only indirectly, although there is a consensus that this inference is based on very reasonable assumptions. If the Unified Model is correct, we infer that the AGNs in which the study of jets is particularly important are the blazars, since we would be observing their axis directly. Besides being candidates to accelerate particles to extreme energies (Chap. 12), jets provide important clues about the environment, such as the magnetic field, density, and other features, maintaining collimation over 50 kpc or more.

While the classification of AGNs according to their radio output goes back to A. Sandage in the 1960s, deeper studies have resulted in a still more refined proposal [6], based not on observational appearance, but rather on the physical model. An examination of the energy output throughout the whole electromagnetic spectrum, its observational biases, and the model behind the emission led to the suggestion that the actual difference between AGNs, based on their X-ray and γ-ray emission, is the presence or absence of a jet [6, 7]. (This is quite analogous to the situation of the supernovae, where the original classification did not say much about the physical origin of the explosions and progenitors.) This suggestion is based on the fact that the radio-quiet AGN class is the dominant population, composed of quasars, while the radio-loud AGNs include a variety of objects emitting non-thermally across the spectrum. Thus, they should be intrinsically different physically. This identification is reinforced by the fact that radio-quiet quasars are actually radio-faint (that is, the ratio of their radio-to-optical emission is low), but they are γ-quiet. The idea is that they lack the physical component that is responsible for the highest energy emission, while the observed thermal component is attributed to the disk.

Padovani and colleagues suggest that the jet is the main difference between the two types, and suggest that the classes should be named "jetted" and "non-jetted" according to whether they have high-energy emission or not. A small set of features, like the presence of a radio excess away from the known far IR–radio correlation, may provide a way to differentiate without the old denomination. Nevertheless, the basic physical picture shown in Fig. 8.5 stands, although the reasons for the presence of a jet that would put the AGN into the "jetted", high-energy emitting category are yet to be clarified.

8.3 AGNs and Structure Formation in the Universe

The presence of quasars in the primordial Universe—the oldest and most distant one detected was formed when the Universe was about 10% of its present size, at a redshift greater than 7.5—means we need to think about the nature of the phenomenon in relation to galaxy formation. According to the most widely accepted model, structure begins to form in the Universe as soon as the growth of density inhomogeneities becomes possible around $z \sim 20$, and as cosmic time proceeds, the number of quasars is observed to grow very rapidly, reaching a peak at $z \geq 2$ and then falling off. Around $z = 2$, almost 10% of galaxies contain quasars, but that number is almost zero today (Fig. 8.6).

As an example, our own galaxy hosts a supermassive black hole in the center Sgr A*, but this does not mean that we are living inside an AGN. This BH is said to be *dormant* or *starved*. There is no steady accretion to power it, and only from

Fig. 8.6 Logarithm of
observed quasar density as a
function of cosmic redshift z,
reflecting the temporal
history of their formation
[9]. After careful analysis by
several authors, it was
concluded that the decline at
earlier times is real, not the
product, for example, of
extinction in a clumpy
Universe. © AAS.
Reproduced with permission

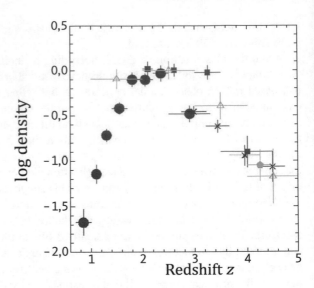

time to time a star or cloud is captured, leading to a brief transient event. So, what
is the actual relationship between quasars and galaxies? We have seen that galaxies
that host them are often difficult to observe, and there are cases where the quasar
is *not* hosted by any galaxy. But if the association is real, it may be that the galaxy
formation process is somehow associated with quasars, as quasars would regulate
the energy released by the star formation that ultimately constitutes the host galaxy.

In fact, when it became possible to study the central regions of galaxies (bulges)
and to estimate the masses of black holes in their centers, an important discovery
was made: the mass of the central black hole is strongly correlated with the velocity
dispersion σ of the stars in the bulges. More precisely, $M_{BH} \propto \sigma^4$ (Fig. 8.7). The
general interpretation of this correlation is that there is a "symbiosis" between the
central black hole and the formation of the inner region, i.e., it looks like the black
hole and the host galaxy formed simultaneously [8].

A very simple idea that lends quantitative support to this M_{BH}–σ correlation can be
obtained by considering the following scenario. Suppose that the radiation pressure
generated in the AGN pushes out on the gas in the center, and suppose that that gas
corresponds to a fraction f of the mass of the bulge, i.e., $M_{gas} = f M_{bulge}$. Using
(7.9) and balancing the pressure force with the gravitational attraction on the gas, we
can write

$$\frac{L_E}{c} = \frac{G M_{bulge} f M_{bulge}}{R^2} . \tag{8.5}$$

One possibility now is to assume that the inner part of the radius R in the bulge corresponds to an isothermal sphere, where the mass and the velocity dispersion of the
"particles" obey the relation $M_{bulge} = 2R\sigma^2/G$. Thus, $M_{bulge}/R = 2\sigma^2/G$, and substituting this in (8.5), we obtain $L_E/c = Gf(2\sigma^2/G)^2$. Using $L_E = 4\pi G M_{BH} m_p/\sigma_T$
from (7.9), we have finally

Fig. 8.7 Correlation
between the black hole mass
M_{BH} and the stellar velocity
dispersion σ in galactic
bulges [10]. Note that this
relationship extends to the
smallest masses of
extragalactic black holes
already detected, less than
$10^5 M_{\odot}$. © AAS.
Reproduced with permission

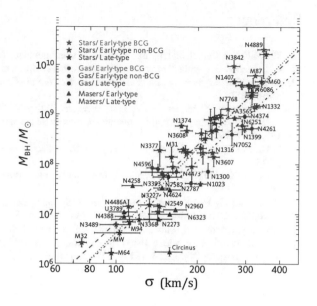

$$M_{\text{BH}} = \frac{\sigma_{\text{T}} f \sigma^4}{\pi G^2 m_{\text{p}}} \, . \tag{8.6}$$

Although it seems a plausible explanation, (8.6) has been criticized. For example, the inner bulge may not correspond to an isothermal sphere and it is quite possible that there are other factors besides the radiation pressure that contribute to the observation that $M_{\text{BH}} \propto \sigma^4$ in Fig. 8.7.

In summary, the largest accreting systems in the Universe, powered by supermassive black holes, formed and evolved from rather high redshifts, greater than 5, and there are even some with $z > 7$. This means that the Universe formed these monster black holes in a relatively short period of time after the Big Bang, something that until recently was considered impossible, but which already seems viable by direct collapse of the gas under the primordial conditions, at least up to the scale of $10^6 M_{\odot}$. The black hole then feeds for billions of years from the ambient accretion, until $\dot{M}$ declines sharply in the local Universe, and the supermassive black hole begins to "starve", like the black hole in Sgr A* at the center of our galaxy, becoming latent, i.e., no longer offering strong signs of its presence. It is clear that the AGN will "vanish" as soon as there is no longer any ambient gas, and this is associated with the evolution of the host galaxy. Otherwise we would observe many nearby quasars still "in operation", and this is not the case. It is hard to obtain accurate statistics for the AGN population, one reason being that the presence of dust makes observation difficult, and may still be hiding a fraction of them [7].

References

1. V.M. Slipher, Radial velocity observations of spiral nebulae. The Observatory **40**, 304 (1917)
2. E.P. Hubble, Extragalactic nebulae. Astrophys. J. **64**, 321 (1926)
3. B. Peterson, *An Introduction to Active Galactic Nuclei* (Cambridge University Press, New York, 1997)
4. V. Beckmann, C. Shrader, *Active Galactic Nuclei* (Wiley-VCH, New York, 2012)
5. C.M. Urry, P. Padovani, Unified schemes for radio-loud active galactic nuclei. ASP Conf. Proc. **107**, 803 (1995)
6. P. Padovani, On the two main classes of active galactic nuclei. Nature Astronomy **1**, 0194 (2017)
7. P. Padovani et al., Active galactic nuclei: what's in a name? Astron. Astrophys. Rev. **25**, 2 (2017)
8. H. Spinrad, *Galaxy Formation and Evolution* (Springer, Berlin, 2005)
9. P. Madau, F. Haardt, M.J. Rees, Radiative transfer in a clumpy Universe. III. The nature of cosmological ionizing sources. Astrophys. J. **514**, 648 (1999). https://doi.org/10.1086/306975
10. N.J. McConnell, C.-P. Ma, Revisiting the scaling relations of black hole masses and host galaxy properties. Astrophys. J. **764**, 184 (2013). https://doi.org/10.1088/0004-637X/764/2/184

Chapter 9
Neutrino Astrophysics

9.1 Neutrinos and Their Detection

The construction of a viable model for the weak interactions and the subsequent discovery of neutrinos postulated by W. Pauli was presented in Chap. 1. We saw that conversions between neutrons and protons—beta decay and inverse beta decay—involve the emission of neutrinos and antineutrinos. As these processes are common in a variety of high energy situations, efforts were made to detect and use them for studies in Astrophysics. Although typical neutrino cross-sections are 20 orders of magnitude smaller than those of electromagnetic interactions, we will see that this possibility, raised in the early 20th century, led to the current success of the nascent neutrino Astrophysics [1].

Several decades after the original Pauli hypothesis, the existence of 3 types of neutrinos (and their respective antineutrinos) had been confirmed in the laboratory, with one for each generation of the Standard Model: ν_e, ν_μ, and ν_τ. This became relevant for our view of the neutrino detection problem: since only protons and neutrons are involved at low energies, we should expect to detect mainly electron neutrinos and antineutrinos (ν_e and $\bar{\nu}_e$). This is no longer true if the temperatures and densities are very high, in which case all three types may be emitted. We will see that this is the case in the gravitational collapse of supernovae.

The fundamental quantity we need to know when trying to detect neutrinos is the cross-section, which is a measure of the effective area over which a neutrino interacts with other matter matter. The phenomena of beta and inverse beta decay suggest that appropriate "targets" should be protons and neutrons, although electron scattering may also be relevant.

Let us suppose that a neutrino collides with a fixed target, for example, a proton. Within the black "dot" of Fig. 9.1, the weak interaction, which is actually mediated by one of the $W^\pm$ bosons of Chap. 1, contributes a factor of G_F^2. Then, since $G_F \propto$ (energy)$^{-2}$, and we have already seen in Chap. 1 that, when we throw a projectile against a target, the length probed satisfies (length) = (energy)$^{-1}$ as in (1.3), the only

© The Author(s), under exclusive license to Springer Nature Switzerland AG 2022
J. E. Horvath, *High-Energy Astrophysics*, Undergraduate Lecture Notes in Physics,
https://doi.org/10.1007/978-3-030-92159-0_9

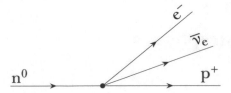

Fig. 9.1 Neutron decay in Fermi's phenomenological theory. The process of emitting a W$^-$ boson with subsequent decay of the latter into e$^-$ + $\bar{v}_e$ is "summarized" by the black dot

way to construct a cross-section with dimensions (length)^2 is to multiply G_F^2 by a quantity with dimensions (energy)^2. In the case of a neutrino incident on a target, the only quantity with these dimensions is the square of the energy in the center of mass, called S for historical reasons. We can show that $S = -(p_v + p_p)^2$, where p_v and p_p are the momenta of the neutrino and proton, respectively. Since the neutrinos are always ultra-relativistic, $p_v = E_v/c$, while for a fixed target $p_p \sim 0$. Now, the energy of the neutrino and the mass of the proton are the relevant quantities. In the low-energy limit $S \propto E_v^2$ and in the high-energy limit $S \propto 2E_v m_p$, both limits being obtained from the general expression for S. Therefore, the total cross-section behaves as [2]

$$\sigma_{vp}|_{\text{low } E} \sim \sigma_0 \left(\frac{E_v}{E_0} \right)^2 , \tag{9.1}$$

$$\sigma_{vp}|_{\text{high } E} \sim \sigma_0 \left(\frac{E_v}{E_0} \right) , \tag{9.2}$$

where $\sigma_0 \approx 10^{-44}$ cm^2, or almost 20 orders of magnitude smaller than the electromagnetic cross-section given in (2.6). The *millibarn* unit $\equiv 10^{-27}$ cm^2 is normally used in scattering problems, but even this figure is actually very large for the purposes of neutrino Astrophysics. This justifies the name "weak interactions" and is used to design experiments where it is important to know beforehand that the cross-section, and with it the rate of events, will *increase* with increasing energy of the incident neutrinos. The more energetic neutrinos are the ones that have the greater probability of interacting.

One can identify two basic types of interaction that can lead to detection: the first is an interaction in which a neutrino provokes a reaction, and with it a change in some chemical element. An example of this type of reaction is

$$v_e + {}^{37}\text{Cl} \rightarrow e^- + {}^{37}\text{Ar} . \tag{9.3}$$

Here, a weak interaction converts a neutron bound to the chlorine nucleus into a proton, and thus the "daughter" atom after the reaction is chemically different and can in principle be separated. The ^{37}Ar count contains information about the number of neutrinos that have reacted. The second type would be a scattering reaction like

$$v_e + e^- \rightarrow v_e + e^- . \tag{9.4}$$

Fig. 9.2 Logarithm of the expected cross-section for neutrinos of energy E_ν (*horizontal axis*). Sources are indicated for each range. Note that the expected flux is not reported, although it is a crucial to know this when assessing the feasibility of an experiment. Adapted from [3]

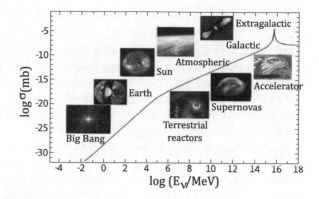

Here it is the charged particle (electron) that can be detected, for example, by the Čerenkov radiation it produces when travelling in water or another liquid. The energy and other parameters of the neutrinos can be obtained by measuring this radiation.

Figure 9.2 shows a realistic estimate of the neutrino cross-section expected from various sources for which the energy E_ν can be reasonably well estimated. From the "relic" Big Bang neutrinos (analogous to the photons in cosmic background radiation) to the end of the spectrum, including a possible resonance due to the *finite mass* of the neutrino (see below), the predictions yield very low values, albeit increasing with energy. In the figure we can observe the change in slope corresponding to the "low energy" $\rightarrow$ "high energy" transition between the limits of (9.1) and (9.2) around $E_0 \approx 10^5$ MeV.

From elementary kinetic theory we can calculate the rate of events in a situation like the incidence of neutrinos onto nucleons, which constitutes a typical fixed target experiment. If we consider an incident neutrino beam with density n_ν covering an area A, the flux onto the targets will be $\Phi = n_\nu A c$. These neutrinos focus on the target, where the density of nucleons is n_N and the thickness is l. The number of events will be proportional to the flux and the cross-section σ, so the total number of events expected per unit time is

$$N_t = \sigma n_\nu v_N A c l . \tag{9.5}$$

Now, a source can produce a continuous flux (constant in time) or a pulsed flux (limited duration). The Sun corresponds to the first type of source, where one can integrate over long times to calculate fluxes and other parameters of the incident neutrinos and the source. The collapse supernovae discussed in Chap. 5 are of the second type: they produce a "burst" of neutrinos in a few seconds with an enormous luminosity, after which the flux drops to undetectable levels. We will discuss these two cases below, as they correspond to the two best known sources among those detected and studied.

9.2   Neutrino Sources: Solar Neutrinos

After World War II, detection techniques and related technologies had advanced to the point where it became feasible to think about experiments to detect neutrinos directly. Theoretical ideas had also been consolidated, and astrophysical estimates had improved greatly, converging to more stable values with a consequent gain in confidence on the part of experimental physicists who were contemplating building detectors.

As discussed in Chap. 4, the basic nuclear reactions of the p–p cycle lead to the production of neutrinos in several of their stages, either with fixed energies ("monoenergetic") or distributions that have a maximum of energy due to kinematic restrictions ("continuous"). Figure 9.3 shows the stages of the reactions where neutrinos are emitted and escape from the Sun.

Given a complete model of the Sun, one can also calculate the fluxes Φ_i emitted by each reaction, in addition to the total flux [4]. Adding all fluxes shown in Fig. 9.4, the total is approximately 7×10^{10} neutrinos $cm^2 s^{-1}$. This number is gigantic: more than *100 billion* neutrinos pass through a human hand held out toward the Sun every second! But the tiny cross-section is responsible for essentially all of them going straight through, without interacting with the matter in the hand. Given these facts, the obvious question is: how could one possibly detect these neutrinos in order to measure the individual spectra of Fig. 9.4 and the total emission of the Sun?

Fig. 9.3 Neutrinos emitted in each reaction stage of the p–p cycle, together with the relevant energies

$$p+p \rightarrow {}^2H + e^+ + \nu_e \qquad E_\nu < 0.42\, MeV$$

$$p+ e^- +p \rightarrow {}^2H + \nu_e \qquad E_\nu = 1.44\, MeV$$

$$e+ {}^7Be \rightarrow {}^7Li + \nu_e \qquad E_\nu = 0.86\, MeV\ (90\%)$$

$${}^8B \rightarrow {}^8B^* + e^+ + \nu_e \qquad E_\nu < 15\, MeV$$

$${}^3He + p \rightarrow {}^4He + e^+ + \nu_e \qquad E_\nu < 18.8\, MeV$$

Fig. 9.4 Predictions for neutrino fluxes and spectra from the Sun. Note the estimated errors in brackets. Figure 1 from [6]

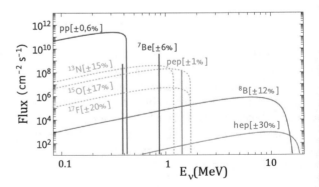

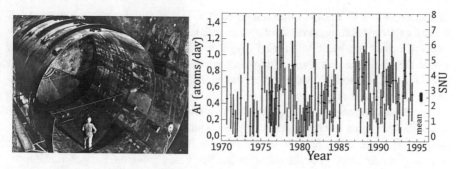

Fig. 9.5 Davies' experiment at the Homestake mine (*left*) and the results obtained there up to 1994 (*right*) [7] © AAS. Reproduced with permission. The *short vertical segment* on the right is the average of these 24 years of operation, equal to about 1/3 of the theoretical expectation

One of the pioneers in this area was Ray Davies Jr., who was the first to consider building an experiment that could measure solar neutrinos. Davies relied on the increasingly sophisticated calculations of fluxes due to John Bahcall and others. An example of the latest predictions is shown in Fig. 9.4. Note that the solar neutrino unit (SNU) was defined for convenience: it is equal to the neutrino flux producing 10^{-36} captures per target atom per second.

Figure 9.5 shows that, year after year, the experimental count fell short of the predictions of solar models. Calculations indicated that some 8 SNU should be measured due to neutrinos with energies above 0.814 MeV, which was the detection threshold that would make it possible to see the monoenergetic flux of ^{7}Be neutrinos. However, the average of the measurements taken in the 40 years of operation was 2.56 ± 0.16 (systematic) ± 0.16 (statistical) SNU. Evidently, this contradiction between theory and experiment cried out for an explanation.

A new series of experiments was planned and executed to clarify the situation. The most important ones, sensitive to low-energy neutrinos, were the SAGE and GALLEX collaborations, which used gallium tanks to measure neutrinos produced in reactions such as

$$\nu_e + {}^{71}\text{Ga} \rightarrow e^- + {}^{71}\text{Ge} . \tag{9.6}$$

Because of the characteristics of gallium in the target compound ($GaCl_3$–HCl), the detection threshold ($E_\nu = 0.42$ MeV) is much lower than the one for ^{37}Cl in (9.3). Thus, the neutrinos of the first p–p reaction, i.e., $p + p \rightarrow d + e^+ + \nu_e$, could be detected with energies $E_\nu > 0.42$ MeV. For the first time, the SAGE and GALLEX experiments were able to see one of the main branches of the p–p chain responsible for most of the energy production in the solar interior, instead of the single energy neutrinos from the beryllium reaction (Fig. 9.4). In fact, the prediction of theoretical models was 69.6 SNU, an order of magnitude greater than the prediction for beryllium neutrinos.

Measurements were made for several years to finally announce that the SAGE results observed a fraction $0.517^{+0.042}_{-0.044}$ (systematic) $^{+0.055}_{-0.053}$ (statistical) of the prediction, while GALLEX communicated $0.601^{+0.059}_{-0.060}$ (total errors). Not only did the two experiments have consistent results with slightly more than half of the prediction, but as they had been "calibrated" by observing neutrinos from nuclear reactors, whose flux was precisely known and where it was established that they could pick up 0.95 ± 0.05 of the total, it became clear that something was wrong with the neutrinos arriving from the Sun [8].

Supplementing these experiments, other collaborations performed neutrino measurements at higher energies. Although less numerous than neutrinos produced in the p–p reaction, these provide a key element for assembling the puzzle. Before resuming our description of the quest for the Sun's *total* neutrino budget, it is important to note that measurements of *particular* neutrino rates can tell us much about the fusion processes going on inside the solar core. We may cite two outstanding examples provided by the BOREXINO Collaboration. The first in 2014 reported measurement of the p–p reaction [9], the fundamental process inside the Sun, giving excellent agreement with solar model predictions. And very recently, the same group reported an accurate measurement of neutrinos produced in the CNO cycle inside the Sun (Fig. 4.10), finding a rate corresponding to around 1% of the whole energy budget [10], but expected to grow rapidly and dominate energy generation in the upper Main Sequence $M \geq 2M_\odot$. It is important to note the degree of accuracy and sensitivity of experiments like this, which allow us to "see" the interior of the Sun and confirm to some extent the nuclear Physics inputs in our theories of stellar structure and evolution.

The most relevant experiments are shown in Fig. 9.6. The first is the Sudbury Neutrino Observatory (SNO), a huge tank that brought together (on loan) the totality of Canada's heavy water. Heavy water has deuterium rather than hydrogen in its composition, and deuterium supports reactions allowing us to detect not only electron neutrinos ν_e, but also neutrinos associated with the muon and the tau. The basic detectable reactions in heavy water are

Fig. 9.6 The Sudbury Neutrino Observatory (SNO) [11] (*left*) and Super-Kamiokande [12] (*right*) experiments. Note the size of these installations, as evidenced by the human operators and the inflatable boat. Credits: MIT group at SNO and Kamioka Observatory, Institute for Cosmic Ray Research, University of Tokyo, respectively

$$v_e + d \rightarrow p + p + e^- \,, \tag{9.7}$$

$$v_x + d \rightarrow p + n + v_x \,, \tag{9.8}$$

$$v_x + e^- \rightarrow v_x + e^- \,, \tag{9.9}$$

where v_x symbolizes either of the neutrinos v_μ and v_τ. These reactions are mediated by the charged W and neutral Z bosons of the weak interactions (Chap. 1), while the third reaction (9.12) is of the elastic scattering-type exemplified by (9.3). The advantage of heavy water is that it sees all neutrinos, whence the total flux and partial contributions can be measured, although only at higher energies.

The experiment shown on the right in Fig. 9.6 is Super-Kamiokande, a cylindrical tank containing 50000 tons of pure water, surrounded by photomultipliers (visible in the image). This configuration is sensitive only to electron neutrinos, but with two important features: it can detect Čerenkov radiation and hence determine the energy in the reactions, while the photomultipliers determine the direction of the incident neutrino with reasonable accuracy.

These experiments could check the feasibility of the simplest ("astrophysical") solution to the solar neutrino problem: a Sun with a slightly cooler temperature at the center. The flux of the ^{7}Be neutrinos, which were the target of the experiment by Davies that revealed the discrepancy, is proportional to a high power of the central solar temperature, viz., $T_{C\odot}^{22}$, so a reduction of only 4% in the central temperature would have been enough to explain the reduced rate originally detected. However, Super-Kamiokande measurements of the neutrino spectrum for the elastic scattering reaction shown in Fig. 9.7 proved that most of the "missing" neutrinos were those of *lower* energy, while a hypothetical decrease in temperature would lead to the

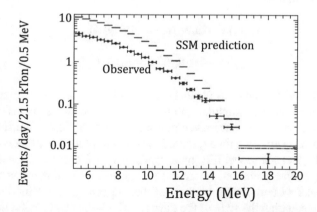

Fig. 9.7 Measurement of the neutrino spectrum [13] compared to the standard solar model (SSM) prediction (*upper discrete bars*), showing the lack of lower energy neutrinos and the gradual convergence between theory and experiment up to around 15 MeV. The so-called "astrophysical" solution was discarded because lowering the temperature of the interior by 4% would produce the opposite behavior: the missing neutrinos would be the ones with the highest energies

immediate disappearance of the higher energy neutrinos (those at the tail of the distribution). Thus, the astrophysical solution was discarded, and suspicions regarding the behavior of the neutrinos themselves increased. As an additional contribution, the SNO measurements showed that the sum of the fluxes of the three neutrino types coincided with the total expected flux, while Super-Kamiokande measured 0.544 ± 0.037 (systematic) ± 0.064 (statistical) for the electron neutrinos, i.e., almost half of the expected flux in the latter was missing.

Following these announcements by Super-Kamiokande and SNO, the whole community agreed that the problem of the lack of neutrinos did not reside in the Sun, but rather in their "disappearance" on their way to the detectors. These suspicions had already been raised by theoretical physicists: while there are reasons for arguing that the photon mass should be zero, there is no fundamental reason for the neutrino mass to be zero. The experimental fact that it needs to be much smaller than the mass of other known particles (electrons, quarks) does not necessarily indicate that it should be zero, although it would be desirable to find some mechanism or reason to explain this hierarchy. Thus, two decades before the first results were obtained for the neutrino flux, B. Pontecorvo [14] suggested that oscillation might be possible between neutrinos of different types, by analogy with the case of kaons, which also oscillate between two states, as a mechanism that would change the emerging flux.

The concept of oscillation between two or more states is not difficult to understand. In Quantum Mechanics, particles are characterized by conveniently labeled states. Technically, a state is a vector, analogous to the well known position vectors in three-dimensional space, although it belongs to a Hilbert space of infinite dimensions. In the same way that any vector has components relative to a set of axes, for example $\mathbf{A} = a\mathbf{x} + b\mathbf{y}$ in two dimensions, any state vector can be expressed in terms of a basis of state vectors, called *eigenstates*. A real electron or muon neutrino (ignoring the tau for the moment) will be a linear combination of mass eigenstates $|\nu_1\rangle$ and $|\nu_2\rangle$ with masses m_1 and m_2:

$$|\nu_e\rangle = \cos\theta_V |\nu_1\rangle + \sin\theta_V |\nu_2\rangle \,, \tag{9.10}$$

$$|\nu_\mu\rangle = -\sin\theta_V |\nu_1\rangle + \cos\theta_V |\nu_2\rangle \,, \tag{9.11}$$

where we have written the coefficients as a function of a single *mixing angle* θ_V, by analogy with a vector in the Euclidean plane. Because the state is a linear combination, a neutrino can change from one mass eigenstate to another. So if a neutrino, say $|\nu_e\rangle$, of energy E propagates for a time t, there is a non-zero probability that it will change to the state $|\nu_\mu\rangle$ and a probability of it remaining in the state $|\nu_e\rangle$. According to the Copenhagen interpretation, where the probabilities are all that can be obtained for any physical system, the probability of this happening is obtained in Quantum Mechanics by squaring the state of the particle. The latter can be calculated as

$$P_{\nu_e}(t) = \left| \langle \nu_e | \nu(t) \rangle \right|^2 = 1 - \sin^2(2\theta_V) \sin^2\left(\frac{\delta m^2 D c^4}{4\hbar c E} \right) \,, \tag{9.12}$$

where $\delta m^2 = m_2^2 - m_1^2$ and D is the distance traveled. We observe that the characteristic length of the oscillation, the distance where the probability reaches unity again, is $L_C = 4\hbar c E / \delta m^2 c^4$, obtained by examining the argument of the second sine function. For there to be a viable fundamental solution to the solar neutrino problem, we must have $L_C < 1$ A.U., since the "disappearance" occurs on the way from the Sun to the Earth. But for neutrinos with $E \approx 1$ MeV, this last condition can be written as $\delta m^2 c^4 \gg 10^{-12}$ eV^2, and the requirement that the mixing angle θ_V should be small (so as not to conflict with observations of neutrinos in reactors and other situations) leads us to discard this proposal for neutrino oscillations in the interplanetary vacuum.

Taking this difficulty into account, Mikheyev and Smirnov [15] showed that, if the oscillation happened in the presence of matter, and not in the vacuum, the probability would have a much higher amplitude and could explain the reduction by 2/3 of the detected flux. According to Mikheyev and Smirnov, if we consider the interaction of the neutrino with matter (electrons and other particles), this leads to an effective mass for these proportional to the electron density [16], and changes the neutrino states. Thus, when neutrinos pass through the Sun, and especially the interface with the "vacuum" immediately outside the surface, they change to other states and disappear from the count when they reach detectors that are insensitive to these rarer types. Hence the importance of total flux measurements made by the SNO experiment: the sum of all the fluxes coincides with expectation, while the partial fluxes of the different types of neutrino change due to the conversion of neutrinos induced by matter.

In order to detect this oscillation, the collaborations built an ingenious experiment (called KamLAND) using neutrinos produced in nuclear reactors, which passed right through the Earth, to be detected in Super-Kamiokande in Japan (Fig. 9.8). As the neutrino flux from the reactor is known, the detection of a much lower value after passing through the Earth confirmed the expected effect. Thus, the experimental determination of the oscillation length L_C in matter could be used to estimate the

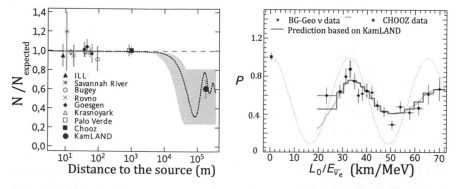

Fig. 9.8 *Left*: Detection of neutrino oscillations on Earth, with the associated determination of the oscillation length [17]. *Right*: Measurements of the oscillation probability compared to predictions (*solid line*) based on previously determined parameters [18]

masses of the participating neutrinos. The results show that these masses are of the order of 10^{-2} eV, and that there are several possibilities for the mixing angle. But the fact is that the Sun does indeed work in the way proposed by our theories of Stellar Evolution, and that the origin of the lack of neutrinos stems from the properties of the neutrinos themselves.

One last point is that it is not yet possible to state whether the mass hierarchy is "normal" ($m_1 < m_2 < m_3$) or "inverted" ($m_1 < m_3 < m_2$). This will be important if we are to build a viable model of massive neutrinos. The Hyper-Kamiokande experiment (in progress [20]) and DUNE (see below) may collect enough data to answer this question in a few years.

9.3 Neutrino Sources: Supernova 1987A

As we saw in Chap. 5, the sequence of events that leads to the implosion and subsequent explosion of massive stars has as its fundamental protagonist the "Fe" core produced during the previous stages. The shock developed by the fall of matter onto the hardened region that exceeds the nuclear saturation density is formed at the *sonic point* and advances later, but may not be the cause of the explosion. This is due to the tremendous losses it suffers on its way to the edge of the core. However, throughout this process, the degenerate core composed of nuclei and electrons is transformed into what we call a *neutron star*. For this to happen, compaction from the original configuration to the final one is mandatory, passing through an intermediate stage (the *proto-neutron star*) where the energy difference that needs to be radiated is emitted by the compact object (Fig. 9.9)

The origin and temporal sequence of the neutrinos in the collapse deserves to be highlighted. Matter is neutronized in the implosion to the left of Fig. 9.9, continuing throughout the collapse, the electrons being "forced" to combine with protons in the

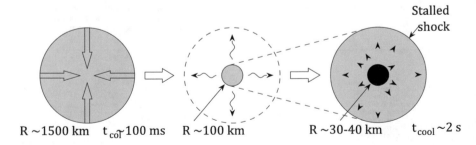

Fig. 9.9 The implosion of the "Fe" core of a massive star is fast and neutronizes all its contents, dissolving the nuclear structure (*left*). A shock forms at the sonic point of the core and advances up to 100–200 km before losses bring it to a halt (*gray region* in the center). The neutron proto-star has a greater radius than it will have when the neutrinos have just left, possibly re-energizing the shock on diffusion scales (1–2 s, *right*). These neutrinos, which carry the binding energy, are detectable on Earth

inverse β reaction

$$p^+ + e^- \rightarrow \nu + n \,. \tag{9.13}$$

This process happens first with protons connected to the nuclei, which we can write as $e^- + A \rightarrow A(Z - 1)$. Electrons are captured and form neutrons, but the nuclei also disappear when the compression is sufficient—which is basically what it means to speak of the nuclear saturation density ρ_0. When the matter reaches this density, the central region suddenly stiffens and the falling envelope "bounces" on it, although we do not know the final sequence of events that leads to ejection of the envelope.

Regardless of the details, the mass of the original iron core, which had a radius of about 1500 km, is compressed into a sphere of tiny radius, about 30–40 km, but with a huge content of thermal energy, dissipated as a direct product of its implosion. Thus, in order to reach a stable configuration, the neutron sphere must radiate (i.e., get rid of) this excess thermal energy. However, instead of doing so by means of photons from the surface, the associated temperatures are so high that it is much more effective to emit neutrinos from the *interior* instead. These neutrinos are *not* the same as those that came out in the "burst" of neutronization while implosion occurred, described in (9.13), but are rather the product of particle–antiparticle annihilations of the type $e^+ + e^- \rightarrow \nu + \bar{\nu}$ and similar processes such as neutron–neutron *bremsstrahlung* $n + n \rightarrow n + n + \nu + \bar{\nu}$ later on. Due to their origin, the neutrinos to be radiated are sometimes called *thermal neutrinos* [21].

As stated above, the neutrinos should take away most of the binding energy, that is, the difference

$$|E_{\text{"Fe"}} - E_{\text{NS}}| \approx 10^{53} \, \text{erg} \,, \tag{9.14}$$

and have to flow in the dense environment that interacts with them, putting them in the *diffusion regime*. It is reasonable to assume that the associated luminosity $L_\nu \propto R_\nu^2 T_\nu^4$, that is, it behaves as a neutrino black body. The fact is that, while the neutrinos leak out in a few seconds, the proto-neutron star adopts its final configuration which will remain unchanged for many millions of years. Thus, purely theoretical considerations point to an burst of neutrinos lasting just a few seconds as the signature of the supernova "core", from which no other signal except this could arrive (apart from gravitational waves, if they are produced; see the next Chapter). In the supernova core, the neutrinos themselves are degenerate, since although the cross-section is very small, the density grows enormously and prevents them from escaping freely. Thus, the neutrinos are expected to be distributed according to the Fermi–Dirac function shown in Fig. 9.10.

In February 1987, a supernova in the Large Magellanic Cloud was identified in an observation beginning at 24.06 UT by astronomer Ian Shelton (Las Campanas Observatory, Chile). According to conventions, it received the name SN 1987A. The progenitor star (identified with the catalog name Sanduleak $-69\,202$) was compared to the predictions of the theory of Stellar Evolution. The star must have remained in the Main Sequence for 10^7 yr, exhausted the hydrogen in its core some 700 000 years ago, passed through a Cepheid stage, and established itself in the helium Main Sequence 650 000 years ago, then leaving it 45 000 years ago and continuing along

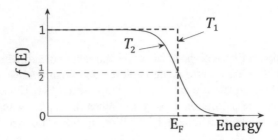

Fig. 9.10 Fermi–Dirac distribution. All possible energy states are filled up to the maximum (the *Fermi level* E_F) for $T_1 = 0$, resulting in the abrupt step shown with *dashed lines*. For a finite T_2, there is room in energy states of order $k_B T_2$ near the Fermi surface (*solid line*) and neutrinos should be emitted with energies of this order, since reactions are only possible if they are not fully degenerate

its accelerated track: carbon ignition 10 000 years ago, neon in 1971, oxygen in 1983, silicon on February 13, 1987, before finally exploding 10 days later. Its effective temperature for most of its existence was over 30 000 K. This was the closest supernova in almost 400 years, and allowed a detailed study of the light curve and other characteristics. However, the most important feature was that, for the first time in history, neutrino detectors were operating and were able to contribute in a remarkable way to our knowledge and understanding of the collapse.

Several detectors in operation at the time (Kamiokande, Baksan, IMB) obtained clear evidence of the existence of a neutrino burst in their data, prior to the optical detection time of the explosion—the neutrinos precede the shock breakout by several hours, because the latter needs this long to reach the surface, while neutrinos escape from the core immediately. However, the distance to the Large Magellanic Cloud conspired against the production of an intense signal, and only 21 neutrinos were definitely associated with the event (Fig. 9.11), since these detections showed

Fig. 9.11 Supernova 1987A neutrinos were registered by IMB (USA) [22] detectors Kamiokande (Japan) [23], and Baksan (Russia) within an interval of about 10 s. The *horizontal time axis* was initiated with the first neutrino of each detector, although there was a relative uncertainty of almost 1 minute between the detectors. The neutrinos detected in the Baksan experiment do not appear in the figure

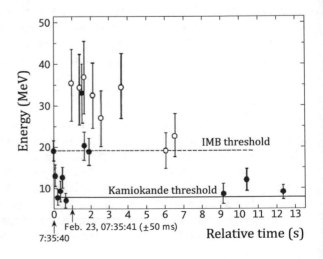

the location inside a cone of aperture a few degrees around the reconstructed axis. This demonstrated that the source of the neutrinos was SN1987A. Thus an attempt was made to "reconstruct" the physical characteristics of the supernova, since the temperature, emitted energy, and other important quantities could be inferred from these data.

Under the hypothesis of neutrino black body emission, there is a simple relationship between the average energy of the emitted neutrinos and the temperature at the source, which can be written as

$$\langle \epsilon_{\text{source}} \rangle = \frac{\int_0^\infty \epsilon^3 f \, d\epsilon}{\int_0^\infty \epsilon^2 f \, d\epsilon} = T \frac{F_3(0)}{F_2(0)} = 3.15T \, , \tag{9.15}$$

where f is the distribution function, assumed to be Fermi-Dirac to a first approximation, and $F_2(0)$ and $F_3(0)$ are called *complete Fermi integrals*, already used extensively when degeneracy is almost total. Because the detectors have different detection efficiencies and detection thresholds H, some functions W are introduced in the integral to calculate the average energy in the detectors. With this caveat, in these detectors we have

$$\langle \epsilon_{\text{det}} \rangle = \frac{\int_H^\infty \epsilon^5 f W \, d\epsilon}{\int_H^\infty \epsilon^4 f W \, d\epsilon} = T \frac{G_5(H/T)}{G_4(H/T)} \, , \tag{9.16}$$

where $G_4(H/T)$ and $G_5(H/T)$ are truncated Fermi functions (also called Fermi functions of the second kind) which are easily calculable numerically. The detector efficiencies of Kamiokande (Japan) and IMB (USA) can be modeled by the functions $W_K = 1 - 4.9 \exp(-\epsilon/3.6 \, \text{MeV})$ and $W_{\text{IMB}} = 1 - 3 \exp(-\epsilon/16 \, \text{MeV})$, and the minimum threshold values $H_K = 7 \, \text{MeV}$ and $H_{\text{IMB}} = 20 \, \text{MeV}$, respectively, i.e., because of their construction it is not possible to detect neutrinos with energies lower than H_K or H_{IMB}.

Now the simplest procedure is to calculate the average energy directly from the data $\langle \epsilon_{\text{det}} \rangle \equiv \Sigma \epsilon_i / N_{\nu \, \text{det}}$, and then solve (9.16) to find the temperature T numerically; this is the temperature of the emitting neutrinosphere T_ν. With this temperature, one can then calculate the average energy at the source $\langle \epsilon_{\text{source}} \rangle$ using (9.15), obtaining the total emitted energy from the source as

$$E_{\text{source}} = 0.77 \times 10^{53} \left(\frac{D}{50 \, \text{kpc}} \right)^2 \left(F_3(0) \frac{G_5}{G_4^2} \right) \frac{N_{\nu \, \text{det}}}{\langle \epsilon_{\text{det}} \rangle M_{\text{det}}} \, \text{erg} \, . \tag{9.17}$$

Another quantity that is directly comparable with the data is the cumulative event rate

$$\frac{d N_{\nu \, \text{det}}}{dt} = 5.2 \times 10^{-8} \left(\frac{50 \, \text{kpc}}{D} \right)^2 M_{\text{det}} C G_4 \left(\frac{T}{\text{MeV}} \right)^5 \, , \tag{9.18}$$

where C is a calculable normalization constant and M_{det} the detector mass.

The results of the predictions for the Kamiokande detectors using (9.18) [23] and IMB [22] came out to be quite reasonable when compared with the actual accumulated event detections, although they depend on the existence of convection inside the proto-neutron star model, which increases the rate by producing a higher effective temperature in the neutrinosphere. Although today the work of modeling events points to a complexity that makes simple prediction difficult due to the effects of instabilities discussed in Chap. 5, the simplest picture agrees to within an order of magnitude with what is observed.

One of the most important parameters determined in this analysis was the neutrinosphere temperature $T_\nu = 4.2^{+1.2}_{-0.8}$ MeV, in good agreement with the simplest models. The temporal evolution of this temperature, and even the possibility of a sequential emission with a "hiatus" of several seconds, cannot be discarded [24], and it may provide evidence for more complex Physics [25]. Finally, the total radiated energy at the source has been compared with the theoretical binding energy for several equations of state proposed for neutron matter, finding consistency, but also leaving undetermined the kind of composition that could have led to the detected events. Note, however, that (9.17) only takes into account the detection of electron antineutrinos, since the electron neutrinos and the muon and tau pairs have very small cross-sections. Thus, the result of (9.17) is usually multiplied by 6 to estimate the total energy carried equally among the 6 types of neutrino produced (e, μ, and τ neutrinos and their respective antineutrinos).

The total radiated energy was $2.5 \pm 1 \times 10^{53}$ erg [26], assuming the formation of a neutron star of mass $1.4 M_\odot$, which may be underestimated, as we saw in Chap. 6. The range of measured masses of neutron stars is today much wider thanks to the measurements of several systems, something that was not known in 1987 (Chap. 5).

The explosion of SN1987A was a fundamental event for neutrino Astronomy, since until then only solar neutrinos had been unequivocally detected. As we have seen, the supernova produced important information about one of the most extreme phenomena in the present Universe. Had a supernova happened inside our own galaxy at a distance $D \sim 1$ kpc, it would have allowed the experiments to register some $20 \times (50 \text{ kpc})^2$ neutrinos, something in the range of 1000–2000, with enormous gains in understanding the mechanism of gravitational collapse. It should be noted that, although it allowed an insight into many fundamental details of the theory underlying this kind of collapse and explosion, we were unable to identify the explosion mechanism itself. And despite numerous subsequent observation attempts, the compact object that was born in the SN 1987A explosion which resulted in the source of the observed neutrinos, i.e., the transformation of the iron core into a neutron star, was never observed, adding a dose of mystery to this problem, in fact leading some researchers to suggest the late formation of a black hole. However, there are recent good indications that a "hot spot" lies in the middle of the SN 1987A remnant, as suggested by ALMA, NuSTAR, and Chandra data. In particular, a non-thermal emission in the range 10–20 keV has been reported by Greco et al. [27] and tentatively associated with a pulsar-wind nebula powered by a young object. If confirmed, this will be the first fully recorded birth of a neutron star, from the explosion itself to infancy.

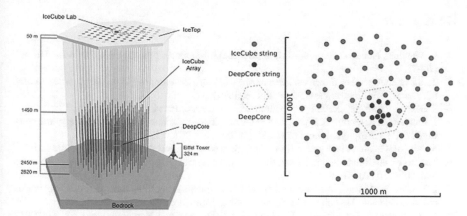

Fig. 9.12 IceCube Čerenkov detector located in Antarctica, which has been operating since 2010. The strings are buried right through the ice to the bedrock, almost 3000 m below. The array itself is located at 1450 m and below. DeepCore's strings are much closer and there are high efficiency photomultipliers inserted in the central part [28]. Credit: Felipe Pedreros, IceCube/NSF

Finally, we remark that new facilities are being built and operated for the observation of astrophysical, cosmological, and geophysical neutrinos, the aim being to investigate processes that end up producing neutrinos in the final state and also to better understand the neutrino masses and the Physics of their interactions.

One good example of these advanced facilities is the IceCube Neutrino Observatory (Fig. 9.12), located near the South Pole. The pristine Antarctic ice generates Čerenkov radiation when neutrinos pass through it, and strings of photomultipliers detect this radiation as described in Chap. 2. This experimental array detected neutrinos with PeV energies and can seek point sources and temporal coincidences with other phenomena (GRBs, for example).

Another important initiative devised to study the Physics of neutrinos is the Deep Underground Neutrino Experiment (DUNE) in the USA. The experiment will use a neutrino beam generated by the Fermilab proton accelerator facility and measure particle interaction near the source, while a second detector located at a distance of 1300 km in a South Dakota underground site (Sanford Underground Research Facility) will measure the beam again to determine, among other things, the neutrino masses and hierarchy (that is, the relation between the masses of the three known flavors). It is clear that neutrino research will remain a key part of high energy Astrophysics for years to come.

References

1. E. Waxman, Neutrino Astrophysics: A new tool for exploring the Universe. Science **315**, 63 (2007)
2. T.D. Lee, *Symmetries, Asymmetries, and the World of Particles (Jessie & John Danz Lectures)* (University Washington Press, Washington, 1987)
3. J. Bahcall, *Neutrino Astrophysics* (Cambridge University Press, Cambridge, UK, 1989)
4. J. Bahcall, A. Serenelli, S. Basu, New solar opacities, abundances, helioseismology, and neutrino fluxes. Astrophys. J. Lett. **621**, L85 (2005)
5. N. Vinyoles et al., A new generation of standard solar models. Astrophys. J. **835**, 202 (2017)
6. The Borexino Collaboration, Comprehensive measurement of pp-chain solar neutrinos. Nature **562**, 505–510 (2018)
7. B.T. Cleveland et al., Measurement of the solar electron neutrino flux with the homestake chlorine detector. Astrophys. J. **496**, 505 (1998). https://doi.org/10.1086/305343
8. M.F. Altmann, R. Mossbauer, L.J.N. Oberauer, Solar neutrinos. Rep. Prog. Phys. **64**, 97 (2001)
9. BOREXINO Collaboration, Neutrinos from the primary proton–proton fusion process in the Sun. Nature **512**, 383 (2014)
10. BOREXINO Collaboration, Experimental evidence of neutrinos produced in the CNO fusion cycle in the Sun. Nature **587**, 577 (2020)
11. http://web.mit.edu/josephf/www/nudm/SNO.html
12. http://www-sk.icrr.u-tokyo.ac.jp/index-e.html
13. J. Hosaka et al., Three flavor neutrino oscillation analysis of atmospheric neutrinos in Super-Kamiokande. Phys. Rev. D **74**, 032002 (2006)
14. B. Pontecorvo, Reviews of topical problems: The neutrino and its role in Astrophysics. Sov. Phys. Uspekhi **6**, 1 (1963)
15. S.P. Mikheyev, AYu. Smirnov, Resonance enhancement of oscillations in matter and solar neutrino spectroscopy. Yad. Fiz. **42**, 1441 (1985)
16. L. Wolfenstein, Neutrino oscillations in matter. Phys. Rev. D **17**, 2369 (1978)
17. K. Eguchi et al., (KamLAND Collaboration), First results from KamLAND: Evidence for reactor antineutrino disappearance. Phys. Rev. Lett. **90**, 021802 (2003)
18. S. Abe et al., (KamLAND Collaboration), Precision measurement of neutrino oscillation parameters with KamLAND. Phys. Rev. Lett. **100**, 221803 (2008)
19. K. Ichimura, *Recent results from KamLAND*, in *Proceedings of the 34th International Conference in High Energy Physics (ICHEP08), Philadelphia, 2008, eConf C080730* (2008). Available at https://arxiv.org/abs/0810.3448
20. http://www.hyper-k.org/en/physics/phys-hierarchy.html
21. A. Burrows, A brief history of the co-evolution of supernova theory with neutrino Physics, in *Proceedings of the Conference on the History of the Neutrino*, eds. J. Dumarchez, M. Cribier, and D. Vignaud. arXiv:1812.05612
22. R.M. Bionta et al., Observation of a neutrino burst in coincidence with supernova 1987A in the Large Magellanic Cloud. Phys. Rev. Lett. **58**, 1494 (1987)
23. K. Hirata et al., Observation of a neutrino burst from the supernova SN1987A. Phys. Rev. Lett. **58**, 1490 (1987)
24. T. Loredo, D.Q. Lamb, Bayesian analysis of neutrinos observed from supernova SN 1987A. Phys. Rev. D **65**, 063002 (2002)
25. O.G. Benvenuto, J.E. Horvath, Evidence for strange matter in supernovae? Phys. Rev. Lett. **63**, 716 (1989)
26. A. Burrows, J.M. Lattimer, Neutrinos from SN 1987A. Astrophys. J. Lett. **318**, L63 (1987)
27. E. Greco et al., Indication of a pulsar wind nebula in the hard X-ray emission from SN 1987A. Astrophys. J. Lett. **908**, L45 (2021)
28. M. Ahlers, K. Helbing, and C. Pérez de Los Heros, Probing particle Physics with IceCube. Eur. J. Phys. C **78**, 924 (2018)

Chapter 10
Gravitational Waves

10.1 Gravitational Radiation: The Basic Physics

Physics has known about and dealt with wave phenomena for a long time. In fact, Physics courses feature a variety of treatments of waves in fluids, electromagnetic waves, and related subjects. On very general grounds we may define a "wave" as a solution of a wave equation. This is not a mere tautology, but an accurate statement that includes many possibilities related to the Physics of wave phenomena. More specifically, a wave equation may contain a variety of terms, but two mandatory ingredients are the second time derivative and the second spatial derivative of some dynamical variable to be determined. Gravitational waves are a true revolution in the understanding of compact objects and gravity itself, a frontier opened a few years ago of high importance in the field. The wave equation treats temporal and spatial coordinates on the same footing, and in its simplest form reads

$$\left(\frac{\partial^2}{\partial t^2} - \frac{1}{v^2} \nabla^2 \right) A = 0 , \tag{10.1}$$

where the symbol ∇^2, called the *Laplace operator*, represents the second-order spatial derivatives, so, for example, $\nabla^2 = \partial^2/\partial x^2$ in Cartesian coordinates in one dimension (Fig. 10.1). The wave of amplitude A in (10.1) propagates with velocity v.

A well-studied case is the electromagnetic wave. We know that the magnetic field **B** and electric field **E** satisfy wave equations in vacuum and propagate with the speed of light c, whenever they are produced by suitably accelerated electric charges. It is always possible to express the resulting electromagnetic radiation in a series of *multipoles*, with the dipole the lowest that produces the waves. The solution for a wave far away from the charges that produced it is depicted in Fig. 10.2.

In the Newtonian theory of gravitation, the gravitational force does not lead to anything like waves. In the Einstenian framework, where the space-time is a kind of dynamical entity affected by the distribution of mass-energy as its source, the motion of the latter produces something quite analogous to the electromagnetic case. The

© The Author(s), under exclusive license to Springer Nature Switzerland AG 2022   203
J. E. Horvath, *High-Energy Astrophysics*, Undergraduate Lecture Notes in Physics,
https://doi.org/10.1007/978-3-030-92159-0_10

Fig. 10.1   Einstein's dilemma: do gravitational waves exist?

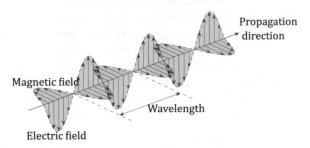

Fig. 10.2   A variable electric field induces a variable magnetic field with the same period, and the two components satisfy a wave equation which propagates by "pushing" each other, with the amplitude varying in the direction perpendicular to the propagation

source of these *gravitational waves* is the acceleration of mass-energy—as we saw in Chap. 1, masses or energies play the role of the "charge" for the gravitational field, because the latter originates from these sources. However, there are some important differences with electromagnetic waves: the electric and magnetic fields are vectors, with a direction and a magnitude. Gravitation is described by a more complex mathematical object, a *tensor*, which may be represented by a matrix in a given coordinate system (actually, scalars and vectors are also tensors, the first of rank 0 and the second of rank 1). Besides this, physically we know that opposite electric charges form a dipole, but there is no such possibility in gravitation. For one thing, mass (or energy) only has one sign. In addition, we can always describe a problem in the center-of-mass system, and because of the equal sign of the mass-energy, the dipole will be zero. We conclude that there is no dipole gravitational radiation, and the lowest mode must be quadrupole.

The usual way of writing a wave equation for gravitational effects is to assume a small (tensor) perturbation that describes the deformations of space-time in a fixed gravitational background represented by the so-called Minkowski tensor $\eta_{\mu\nu}$, which is the solution for the gravitational field in the absence of matter-energy, i.e., we focus on the radiation zone, far from the sources. This perturbation will be called $h_{\mu\nu}$. The full derivation consists in substituting $\eta_{\mu\nu} + h_{\mu\nu}$ into the non-linear Einstein equations and keeping the linear terms. The result is

$$\left(\frac{\partial^2}{\partial t^2} - \frac{1}{v^2} \nabla^2 \right) h_{\mu\nu} = 0 . \tag{10.2}$$

This is a standard calculation and will not be repeated here. The reader may consult [1], for example, for an appraisal of the derivation and its physical discussion, but this is not crucial for our purposes. It is enough to observe that (10.2) has the form of a wave equation for the quantity $h_{\mu\nu}$, giving a deformation that propagates with speed v transversely to the direction of the wave. The identification of v with the speed of light c is almost immediate [2].

Adopting the x-axis as the propagation direction we may write, analogously to the electromagnetic problem, a general factorized form for the amplitude:

$$h_{\mu\nu} = A_{\mu\nu} \exp i k_\alpha x^\alpha , \tag{10.3}$$

where we have written $k_\alpha x^\alpha$ to indicate the scalar product of the 4-wave vector with the direction. In four dimensions, this product has the well-known form $(\omega t/c - \mathbf{k}.\mathbf{x})$. The deformation of space-time $h_{\mu\nu}$, being orthogonal to the propagation, can be further decomposed into two independent modes, conventionally denoted by $+$ and $\times$ (*plus* and *cross*). The associated amplitudes h_+ and $h_\times$ can be combined to yield $A_{\mu\nu}$ according to

$$A_{\mu\nu} = h_+ e_{\mu\nu}^+ + h_\times e_{\mu\nu}^\times . \tag{10.4}$$

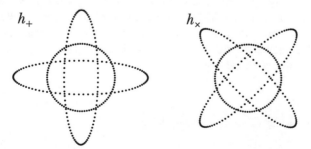

Fig. 10.3 Polarization modes h_+ and $h_\times$. The direction of propagation (x-axis) is perpendicular to the plane of the page. The $(+)$ mode, on the left, "stretches" matter in the vertical direction while it compresses it horizontally, and the $(\times)$ mode does the same, but at an angle of $45°$, when the wave passes. Any deformation $A_{\mu\nu}$ can be fully decomposed into these two modes, provided that the amplitudes h_+ and $h_\times$ are determined

The geometrical interpretation of these two modes is shown in Fig. 10.3.

In 1918, Einstein derived the dynamical equation for the quantity $h_{\mu\nu}$ as sketched above and further showed that this is proportional to the second time derivative of the quadrupole moment Q of a mass-energy distribution, i.e., $h_{\mu\nu} \propto \ddot{Q}$. Therefore, the emitted power (or luminosity) in gravitational waves is proportional to the *third* time derivative of Q :

$$L_{GW} \propto \dddot{Q} . \tag{10.5}$$

However, he doubted the reality of this result for years (Fig. 10.1). There was a lively discussion about whether the perturbations could be made to vanish with an appropriate coordinate change, i.e., about whether the perturbative wave solutions might actually be spurious, stemming from a choice of coordinate system. However, around 1950, the reality of the waves had finally been demonstrated, even though the prospects for detecting them were considered null or at best remote.

To appreciate the actual magnitude of this physical phenomenon, we can assume that the motion of masses at the source occurs on a characteristic timescale τ, roughly a measure of the time a mass takes to relocate within the system. On the other hand, the quadrupole moment is approximately the product of the mass M times the square R^2 of the dimension of the system. Thus, we immediately find the estimate

$$\dddot{Q} \approx \frac{MR^2}{\tau^3} \approx \frac{Mv^2}{\tau} \approx \frac{E_{NS}}{\tau} , \tag{10.6}$$

where v is the speed of the masses in motion and E_{NS} is the energy associated with the non-spherical part of the system. Moreover, for a self-gravitating system, we also have $\tau \approx \sqrt{R^3/GM}$, and with it a characteristic frequency $\nu = 2\pi/\tau = 2\pi f$. We can then estimate the resulting luminosity as

$$L_{GW} \approx \frac{G^4}{c^5} \left(\frac{M}{R}\right)^5 \approx \frac{G}{c^5} \left(\frac{M}{R}\right)^2 \approx \frac{c^5}{G} \left(\frac{R_S}{R}\right)^2 \left(\frac{v}{c}\right)^6 , \tag{10.7}$$

where we have introduced the Schwarzschild radius R_S defined in Chap. 6. From this formula it is evident that the highest luminosities will be produced by compact objects ($R \approx R_S$) moving at relativistic speeds ($v \approx c$). We must now study the kinds of event in actual astrophysical systems that could be the best candidates for the detection of gravitational waves.

10.2   Sources of Gravitational Waves

Having estimated the luminosity of gravitational waves on general grounds, we may attempt to determine which systems and events look most promising for their detection. But this likelihood alone does not say much, because we also need to estimate the amplitude of the perturbation for a source at a distance r for concrete experimental purposes. With the same degree of approximation as before, the value of this dimensionless amplitude of space-time deformation is

$$h = \frac{G}{c^4}\left(\frac{E_{NS}}{r}\right) \approx \frac{G}{c^4}\left(\frac{\epsilon E_k}{r}\right) , \tag{10.8}$$

where ϵ is the fraction of the kinetic energy E_k actually emitted in waves. Numerically, we have

$$h = 10^{-22}\left(\frac{E_{GW}}{10^{-4}M_\odot c^2}\right)^{1/2}\left(\frac{1\,\text{kHz}}{f_{GW}}\right)\left(\frac{\tau}{1\,\text{ms}}\right)^{-1/2}\left(\frac{15\,\text{Mpc}}{r}\right) . \tag{10.9}$$

The choice of a scaling distance $D = 15$ Mpc is not arbitrary: it corresponds to the mean distance to the Virgo galaxy cluster, where there are some 10 000 galaxies. This cluster of galaxies was originally expected to produce several tens of events involving non-spherical gravitational collapse every year (provided that the non-spherical deformation of the source was large enough). It least this was the original expectation of those researchers who began to think about building GW detectors [3]. Although this expectation for non-spherical collapses was not fulfilled, we will now show how binary mergers much further away produced measurable signals once the detectors were up and running.

The estimate in (10.9) implicitly assumes a "burst" of emission of short duration (of ms order) and characteristic frequency of order 1 kHz. However, *continuous* sources can also be considered, with frequencies that change slowly with time. In fact, all the compact binary systems belong to this type, because the orbit produces a variable quadrupole moment. Consider a generic binary system with masses M_1 and M_2 and semi-axis a, such as the one shown in Fig. 10.4. The semi-axis of this system is expressed as $a = a_1 + a_2$ and the reduced mass is $\mu = M_1 M_2/(M_1 + M_2) \equiv M_1 M_2/M$. The only non-zero components of the quadrupole moment of the binary with the orbit located in the xy plane are

$$Q_{xy} = Q_{yx} = \frac{1}{2}\mu a^2 \sin(2\Omega t) , \tag{10.10}$$

where the orbital frequency Ω can be obtained from Kepler's third law $\Omega^2 = GM/a^3$, noting that the frequency of the quadrupole is twice the orbital frequency Ω. There-

Fig. 10.4 Generic compact binary with masses M_1 and M_2, and semi-axis a

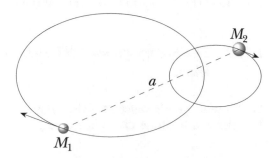

fore, the luminosity is just

$$L_{GW} = -\frac{dE}{dt} = \frac{G}{5c^5} \left(\mu \Omega a^2\right)^2 \left[2\sin^2(2\Omega t) + 2\cos^2(2\Omega t)\right]$$

$$= \frac{32}{5}\frac{G}{c^5}\mu^2 a^4 \Omega^6 = \frac{32}{5}\frac{G^4}{c^5}\frac{\mu^2 M^3}{a^5}. \tag{10.11}$$

We see from this that the emission grows enormously as the semi-axis a shrinks, that is, in the advanced stages before the final merging. The total energy of the system is

$$E = \frac{1}{2}\Omega^2\left(Ma_1^2 + M_2 a_2^2\right) - \frac{GM_1 M_2}{a} = -\frac{1}{2}\frac{G\mu M}{a}. \tag{10.12}$$

As time goes by, gravitational radiation makes the orbit shrink at a rate obtained from

$$\frac{dE}{dt} = -\frac{G\mu M}{2a^2}\frac{da}{dt} \implies \frac{da}{dt} = -\frac{64}{5}\frac{G^3}{c^5}\frac{\mu M}{a^3}, \tag{10.13}$$

where we have once again used Kepler's third law. Since the orbital frequency increases as $3\dot{a}/2a$, we can integrate (10.13) to find the time for the binary system to merge, assuming it starts from an initial semi-axis a_0, with the result

$$\tau = \frac{5}{256}\frac{c^5}{G^3}\frac{a_0^4}{\mu M^4}. \tag{10.14}$$

Finally, using the above expressions, we can estimate the dimensionless amplitude of the resulting gravitational waves as

$$h = 5\times 10^{-22}\left(\frac{M}{2.8M_\odot}\right)^{2/3}\left(\frac{\mu}{0.7M_\odot}\right)\left(\frac{f_{GW}}{100\,\text{Hz}}\right)^{2/3}\left(\frac{15\,\text{Mpc}}{r}\right), \tag{10.15}$$

where the numbers correspond to a symmetric binary system in which each component has mass $1.4M_\odot$, as is commonly assumed for neutron star binary systems (Chap 6). We find that a coalescing binary located in one of the galaxies of the Virgo cluster would produce a signal comparable to the burst in (10.9). However, since many closer binaries are known, the probability of detection, at least indirectly, was immediately considered. This possibility prompted a study by Hulse and Taylor which led to the 1993 Nobel Prize in Physics and which will be described in the following.

10.2.1   The Binary Pulsar PSR 1913+16 and Gravitational Waves

In 1974 was a year in which relativistic Astrophysics gained a major stimulus, when J. Taylor and R. Hulse discovered a very particular binary system: while one of the components is a pulsar, the other is also a compact object with similar mass. Even though pulsations from the companion are not detected, it was finally identified as another neutron star. This identification became possible because the binary was observed with high accuracy, first to determine the mass function $f(M_1, M_2, i)$ defined in (6.38), with $P_{orb} = 2.79 \times 10^4$ s and $GM_\odot/c^3 = 4.925\,490\,947\,\mu$s a constant with dimensions of time. All the quantities on the right-hand side of (6.38) have been very accurately measured, and the pulses are seen as advanced or delayed depending on whether the pulsar is approaching or receding from the observer, so the radial velocity is also known via the Doppler effect. The whole system can be completely determined by modeling the gravitational field of the companion. The orbit is quite eccentric and inclined at about 45° with respect to the line of sight—this is the angle i in (6.38). In summary, we have a very precise "clock" (the pulsar) orbiting a companion with a strong gravitational field, and thus a variety of general relativistic effects allowing an accurate knowledge of the system. A representation of PSR 1913+16 and its companion is shown in Fig. 10.5 [4].

Now, with the help of the expression (10.15), we can immediately obtain the gravitational radiation emitted with frequency $f_{GW} = 1.1 \times 10^{-5}$ Hz and amplitude $h \sim 10^{-23}$. These values happen to be very low and their direct detection would be impossible at the present time. But there is a way around this: since every aspect of the system has been very accurately measured, (10.13) can be used to calculate the rate of change of the semi-axis, and with it the change resulting from the passage of the neutron star through the periastron, which results in a decrease of the orbital period by

$$\Delta P_{orb} = -7.6 \times 10^{-5} \text{ s/yr} , \qquad (10.16)$$

a value much greater than the accuracy attained in the measurements, including errors and uncertainties of various origins. Therefore, by monitoring the binary pulsar, the

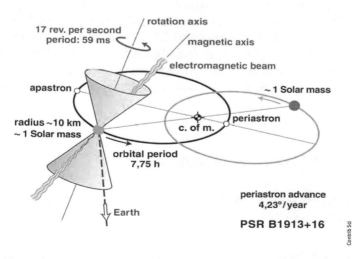

Fig. 10.5 Representation of PSR 1913+16 and its companion, another neutron star. The inferred orbital parameters and masses are indicated [5, Fig. 2]

Fig. 10.6 Comparison between the measured data (*dots*) and the prediction of General Relativity (*full curve*), showing that the cumulative shift of the periastron time after about 20 yr agrees with the former to within about 0.1%, and lies far away from the horizontal line that would indicate no decay at all (obviously disfavored by the data)

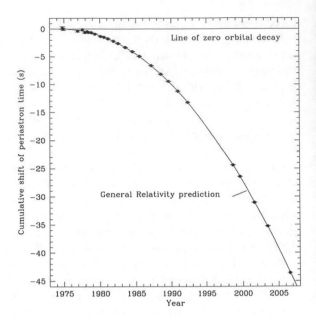

accumulated change can be determined and the result compared with the prediction (10.16), as shown in Fig. 10.6.

As mentioned above, Hulse and Taylor were awarded the Nobel Prize of Physics in 1993 for their discovery of the binary pulsar and the work that demonstrated the agreement of the decay of the orbit with the predictions of General Relativity, indirectly confirming that gravitational radiation is being emitted from the system. Deviations from General Relativity smaller than 0.1% are still possible, showing that there is little room for alternative theories of gravitation, at least with regard to the gravitational radiation emission produced in this kind of binary system. These results were improved by an analysis of the system PSR J0737-3039A/B announced in 2003 [6], in which two pulsars move with an orbital period of 2.4 hours. The decay of the orbit for that system was measured to be in agreement with General Relativity with a precision of around 0.01%, and a few additional relativistic effects were measured for the first time.

10.3 Gravitational Wave Detectors

10.3.1 Interferometers and Resonant Masses: From Dreams to Reality

Now that we have discussed the main sources of gravitational waves and we understand the central idea of space-time deformation a little better, we can present the

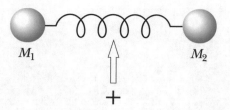

Fig. 10.7 Direct detection of gravitational waves. Two masses M_1 and M_2 joined by a spring react to a gravitational wave travelling perpendicular to the spring with a hypothetical $+$ polarization, like the one on the left in Fig. 10.3

principles of the detectors built since the beginning of the second half of the 20th century, which have turned the observation of gravitational waves into a reality in contemporary science. Basically, there are two ways to detect the passage of these waves directly—recalling that the decay of the orbit of the system PSR 1913+16 is an indirect approach. The first is to measure the deposition of energy transported by the waves in a mass, through a measurement of the mechanical oscillations they induce. The second is to monitor masses that are perturbed by the passage of the waves, but where no deposition of energy is involved. These detectors are called resonant masses and interferometers, respectively.

Consider the situation in Fig. 10.7, showing the simplest possibility for a resonant detector. If the incident wave has proper frequency ω and the two-mass oscillator a natural frequency ω_0, the equation of motion for the amplitude ξ is

$$\ddot{\xi} + \frac{\dot{\xi}}{\chi} + \omega_0^2 \xi = -\frac{1}{2}\omega^2 L h_+ e^{i\omega t} , \qquad (10.17)$$

where χ represents the friction, L is the distance between the masses, and the force on the right-hand side stems from the incident wave, which "shakes" the oscillator. The solution of this equation is well known and reads

$$\xi(t) = -\frac{\omega^2 L h_+}{2(\omega_0^2 - \omega^2 + i\omega/\chi)} e^{i\omega t} , \qquad (10.18)$$

that is, there is a *resonance* when $\omega \approx \omega_0$ and the maximum amplitude is $\xi_{max} = 1/2\omega_0\chi L h_+$. If we are to construct an actual detector, it helps to maximize the product $\omega_0 \chi L$. In this way, besides choosing a material with a high Q factor enabling a clear measurement of the induced oscillations, the detector L must be as large as possible, i.e., the greater the mass, the more energy will be absorbed.

Bearing this in mind, the first attempts to construct a resonant detector in the 1960s used aluminum with a high Q and had a typical proper frequency $\omega_0 \approx 1650$ Hz, reaching a sensitivity of $h \sim 10^{-15}$. This last figure is surprisingly good, equivalent to detecting an amplitude of the order of the proton radius with a bar of length 1.5 m, but it is still insufficient to observe any naturally occurring events. The detectors operated at room temperature and the thermal noise was very difficult to control,

Fig. 10.8 *Left*: Joe Weber working with one of his resonant bars around 1965. Credit: Special Collections and University Archives, University of Maryland Libraries. *Right*: Image of the detector Nautilus, in Italy, one of the bars operated 30 years later, with improvements in the electromechanical transduction, the suspension, and the suppression of thermal noise due to cryogenic operation at $T \ll 1$ K, reaching sensitivities up to $h \sim 10^{-19}$ at the center of the (narrow) operating band. Credit: Frascati Laboratories, INFN

ultimately limiting the sensitivity of the antennae. On several occasions, J. Weber at the University of Maryland (USA), who pioneered these attempts, claimed positive detections that would have implied huge energies channeled into gravitational waves in every putative event [7], but these were never confirmed by other groups. Nevertheless, his pioneering work later led to the construction of advanced detectors with much improved sensitivity (Fig. 10.8).

Only three resonant antennae are running today, since the approval, financing, and operation of the LIGO observatory in USA and Virgo in Italy have somewhat reduced the resources available for this kind of experiment. One of the reasons why interferometers are attractive is that, by their very nature, the resonant masses have good sensitivity around a restricted band near ω_0. Although the latest transducers— the devices that "translate" the mechanical oscillations into an electronic signal—can widen this bandwidth, this amounts to only 10–20 Hz around the central frequency. Therefore, the idea of observing the radiation from binaries at around 100 Hz shifted interest to interferometers, capable of registering from around 10 Hz up to around 1 kHz, albeit with different sensitivities. There are nevertheless plans to continue these experiments with resonant antennae, which can achieve $h > 10^{-20}$ today, but their future depends on several factors which are difficult to evaluate.

The construction of wide-band interferometers, on the other hand, had to overcome several technological challenges to become a reality. Since the basic principle is to monitor the relative positions of masses, mirrors were attached to them and laser light made to reflect off the mirrors. The original light beam is divided into two parts which follow orthogonal paths (Fig. 10.9). Upon their return, the laser wavefronts are made to interfere, and if the mirrors oscillate due to the passage of a gravitational wave, a variable interference pattern is observed at the photodetectors. An important advance,

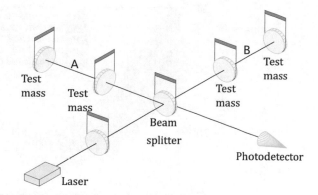

Fig. 10.9 Fabry–Perot interferometer, indicating some of the components. The beams are made to interfere and detected in the photomultiplier

which is in fact what permitted the actual detections (see below) was the presence of two Fabry–Perot cavities, one on each arm of the interferometer. This way, the laser light is made to travel back and forth many times, increasing the effective length of the arm well beyond the actual, physical size of about 4 km, achieving effective lengths greater than 1000 km. This is similar to what is done in the mirror houses of amusement parks, in which images are multiplied using a configuration of parallel mirrors [8].

This interferometric technique can detect extremely tiny oscillations. In fact, the amplitude at the center of the band is $\Delta l = hl \sim 10^{-17}$ cm (!) for an arm length of 4 km. This is equivalent to 0.0001 of a proton diameter. As stated earlier, interferometers are sensitive to all arriving waveforms, but in each range of frequencies the dominant noise that complicates the detection has different origins. Figure 10.10 shows the sensitivity curves as measured and desired for the successive data runs of the experiment known as the Laser Interferometer Gravitational Wave Observatory (LIGO). One can see that the curve improved in each run, allowing better sensitivities, and the collaboration hopes to reach the lower curve with an advanced design. The best frequency, where the interferometer is most sensitive, is around 100 Hz, where a compromise between noise sources is attained. Above that frequency, the curve rises due to the uncertainty in the emission of photons by the laser (called *shot noise*) and the sensitivity gets progressively worse at the highest end. On the other hand, below about 30 Hz, the continual seismic vibrations of the Earth's crust become important, and although they can be attenuated using sophisticated mechanical systems, they nevertheless limit detection at the lowest end. Therefore, the most reasonable strategy is to try to improve performance in measuring events around the center of the band, leaving the high and low frequencies as they are. This has indeed been the general decision made by the LIGO collaboration during the development of the present setup.

Bearing in mind our previous discussion of binaries, we can see from (10.11) that the emitted luminosity in gravitational waves is proportional to the reciprocal of the semi-axis a to the fifth power. This means that the majority of inspiraling binary systems will in general be far from merging and will emit low frequencies (recall Kepler's third law $\Omega^2 = GM/a^3$). A successful detection of these systems using an

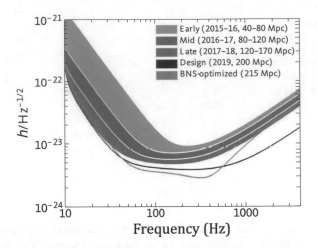

Fig. 10.10 Sensitivity curves of the LIGO experiment [9]. The data run S6, achieving a sensitivity of $h < 10^{-22}$ around 100 Hz, has now been improved thanks to an upgrade of several components. Various "peaks" in the curve due to several mechanical resonances of the system—not fully identified even using the best available numerical simulations—are not shown in this figure. The collaboration operates the interferometers near the "design" goal in black. This has been sufficient to detect several mergers (see text). Credit: B.P. Abbott et al., Living Reviews in Relativity **19**, 1 (2016)

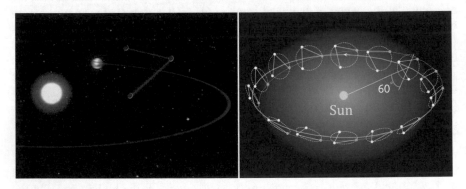

Fig. 10.11 Orbit planned for the eLISA space interferometer, designed to minimize solar perturbations and obtain a wide coverage of the sky. Credit: Cardiff University, Physics and Astronomy Outreach [11]

interferometer would need extremely long arms since the oscillation frequency is inversely proportional to the arm length, and also very low noise perturbation. For these reasons, the best solution seems to be the construction of a space interferometer.

Figure 10.11 shows the main features of the project known as the evolved Laser Interferometer Space Antenna (eLISA) under study by the European Space Agency. One interferometer with a "mother" satellite and two "daughters" set on an equilateral triangle of side about 1 million km orbiting 50 million km from the Earth will be capable of detecting binaries with emission centered around 10^{-2} Hz by monitoring the relative motions of the three spacecraft (Fig. 10.12). This is provided that a range

Fig. 10.12 Sensitivity curves for existing and planned experiments. The *red star* shows the amplitude of the events initially detected by the LIGO–Virgo collaboration (see text). Adapted from Fig. A1 of [12]

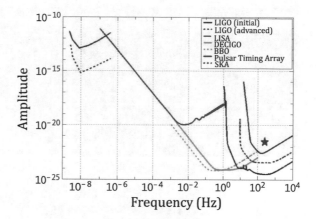

of perturbations like the solar wind can be kept under control. The accuracy required for these measurements involves measuring the position of each satellite with an error of the order of 10^{-14} cm, which seems reasonable with the existing technology. Note that the projected space interferometer, of the Michelson type, will not carry mirrors in the "daughter" satellites, because the laser beam would be too weak to be reflected. The idea is to retransmit the signal from each satellite in active form. At the moment, tests are being conducted for a time horizon beyond 2030 [10]. Other missions and experiments designed to probe the lowest frequencies are shown in Fig. 10.12.

10.4 Detection of Black Hole and Neutron Star Mergers: The Beginning of a New Era

10.4.1 Overture: The Black Hole Merger Event GW150914

The initial operation of the two twin LIGO interferometers and later the French–Italian project Virgo (Fig. 10.13) raised expectations because the initial sensitivity curves already shown in Fig. 10.10 were encouraging for the detection of events, provided that they occurred during data runs with enough intensity. The long-awaited day finally arrived on September 14, 2015, when the two LIGO interferometers detected a simultaneous signal, the first identified as such, and were able to discard a noise origin. Hence began the era of gravitational wave Astrophysics [13]. Figure 10.14 shows the data. The probability of a mere coincidence (random fluctuation) rather than a real event is extremely low, with a confidence level of more than 99.99994%, i.e., a statistical significance greater than 5σ.

Even though there was no doubt that the event originated from a binary merger, a comparison with theoretically simulated templates of the waveform had to be carried out to extract the parameters characterizing the individual members and the event in general. The accepted interpretation is that the event was produced by a merger of two black holes of stellar mass, giving rise to a more massive object and radiating the

Fig. 10.13 Aerial views of the LIGO interferometers in Livingston (Louisiana) and Hanford (Washington), *top left* and *top right*, respectively. *Lower*: VIRGO interferometer in Cascina (Italy). Credits: Caltech/MIT/LIGO Lab and Virgo Collaboration

Fig. 10.14 Signals recorded in Hanford (*upper*) and Livingston (*lower*), coinciding with the second merger of two black holes GW151226, quite similar to the first GW150914. The approach stage (inspiral), the merger (occurring at time zero), and the vibration of the resulting object (ringdown) after the merger are all clearly visible. Credit: LIGO Collaboration

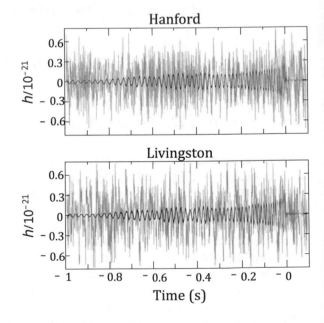

excess energy away mainly in the form of gravitational waves. This is the conclusion from the analysis of the gravitational wave form in the approaching stage, yielding the quantity known as the *chirp mass* $\mathcal{M}$, which is a combination of the individual masses M_1 and M_2 of the form

$$M = \frac{(M_1 M_2)^{3/5}}{(M_1 + M_2)^{1/5}} = \frac{c^3}{G} \left(\frac{5\dot{f}}{96\pi^{8/3} f^{11/3}} \right)^{3/5} . \tag{10.19}$$

From the data for this first event GW150914, $M \approx 30 M_\odot$. On the other hand, the sum of the masses must satisfy $M_1 + M_2 > 70 M_\odot$ in order to accurately match the post-merging stage. The individual masses came to $M_1 = 36^{+5}_{-4} M_\odot$ and $M_2 = (29 \pm 4) M_\odot$. Only two black holes could have produced this collision. Later on, a few similar events were detected (for example, GW151226 in Fig. 10.14), and it is currently possible to monitor a population of binary black holes merging at cosmological distances.

From the same analysis of the waveform it is inferred that, after the merger GW150914, the product had a mass of $M_f = (62 \pm 4) M_\odot$, and therefore $(3 \pm 0.5) M_\odot$ was radiated in gravitational waves. Finally, it was concluded that the spin of the resulting black hole reached about 2/3 of the maximum possible value, a likely consequence of the partial transfer of the orbital angular momentum of the progenitor black holes. A Schwarzschild black hole without spin would not match the observed waveform. The spatial velocity of the colliding black holes near the time of the merger was greater than $0.5c$. Finally, from the observed luminosity it was possible to calculate the luminosity distance to the event, with the result 440^{+160}_{-180} Mpc, that is, a cosmological distance scale. Even if this was the first event of this type registered in the history of modern Astrophysics, the results showed that the visibility of this class of mergers reaches a considerable fraction of the whole observable Universe.

Given the continuity of the observations, the detection of events similar to GW150914 was not surprising. Figure 10.15 depicts the set of such events observed by the end of 2020. Black hole mergers were not expected to be frequent, and in fact the strongest bet of the community was first to detect the merger of two neutron stars. However, it is also worth noting that the individual masses of the black holes in merging systems are much higher than those inferred for these objects in nearby binary systems (see Fig. 6.29). In fact, the individual colliding masses are on average greater than $20 M_\odot$. This fact suggests that, in the early Universe, the stars that originated this population should have high masses. This is consistent with the predictions of the theory of Stellar Evolution, which indicate the formation of stars with $M \geq 100 M_\odot$ whenever metallicity is very low. This observation shows how important it was to open this new window on the Universe, and everything started with the detection of GW150914 discussed here.

10.4.2 The Aftermath: A Merger of Neutron Stars in GW170817

After the confirmation of the events discussed above, it became legitimate to ask where were the "most likely" events expected in the community, that is, the mergers

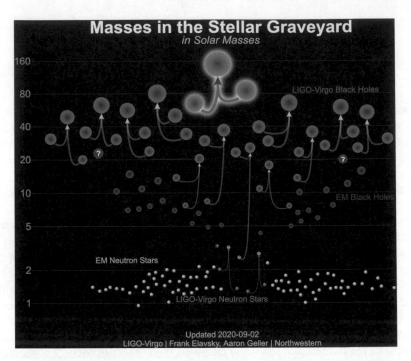

Fig. 10.15 Map of the events detected by the LIGO/Virgo collaboration by the end of 2018. The first announced GW150914 is the first in blue to the left. Note the difference with the typical masses of black holes in binary systems in our galaxy (*purple dots*). Credit: LIGO-Virgo/Frank Elavsky, Aaron Geller/Northwestern

occurring in neutron star–neutron star binaries. Prior to the first science runs, the vast majority of astrophysicists strongly believed that their detection was imminent, but this is not what happened. However, almost two years after the detection of GW150914, an event showing the features of a NS–NS merger was finally recorded [14]. If judged from the point of view of the far-reaching consequences of simultaneous multiple observations, it was worth waiting for, as we shall see.

On August 17, 2017, all three interferometers LIGO/Virgo simultaneously detected a strong signal. In addition, the gamma-ray satellites FERMI and INTEGRAL saw a short gamma burst, lasting around 2 s and located at the periphery of the galaxy NGC 4993. From the chirp mass waveform measurements, it was inferred that two compact objects with total mass $M_1 + M_2 = 2.74^{+0.04}_{-0.01} M_\odot$ had collided and merged, although their individual masses could not be determined. However, using the same waveform and numerical modeling, the components were found to lie in the intervals $M_1 = [1.36M_\odot, 1.6M_\odot]$ and $M_2 = [1.17M_\odot, 1.36M_\odot]$, ignoring the effects of spin (the waveform does not indicate any evidence for non-zero spin). Note that, even if though is not mandatory, it seems likely that the two individual masses are actually equal, with a value of $1.36M_\odot$ (although this is not guaranteed and has been disputed [15, 16]). The same type of analysis showed that the event is consistent with a merger of two neutron stars, and not with another type of event, such as a NS–BH merger.

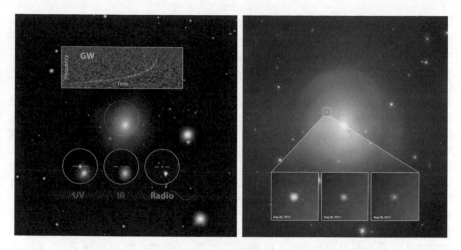

Fig. 10.16 *Left*: Detected signal from the event GW170817 (*inset*) and the localization of the same in gamma-rays (*center*), accompanied by detections in the UV, infrared, and radio (*lower*) achieved by more than 60 instruments. Credit: Robert Hurt (Caltech/IPAC), Mansi Kasliwal (Caltech), Gregg Hallinan (Caltech), Phil Evans (NASA) and the GROWTH collaboration. *Right*: Optical images obtained by the Hubble Space Telescope on August 22, 26, and 28, 2017, showing the fading optical magnitude of the associated transient. Credit: NASA and ESA: A. Levan (U. Warwick), N. Tanvir (U. Leicester), and A. Fruchter and O. Fox (STScI)

The event GW170817 provided an extraordinary benchmark for gravitational wave detection, but also for other areas of Astrophysics. We mentioned that the FERMI satellite detected a short gamma-ray burst in coincidence (see the next Chapter), 1.74 ± 0.05 s after the "zero" of the gravitational merger, with a duration of about 2 s, thus confirming that neutron star mergers are sources of "short" GRBs, although the intensity of the burst was a factor of 1000 lower than expected, possibly due to the angle with the line of sight [17]. Other instruments monitored the burst at lower frequencies, all the way down to the optical and infrared (Fig. 10.16). The first of these detections was performed by the collaboration known as the *Swope Supernova Survey*, almost 11 h after the merge time shown by the gravitational waves. It is interesting that absorption lines in the infrared were associated with the formation of lanthanides producing high opacity in the ejecta [18], and blamed for the temporal behavior of the light curve. The late glow has been called a *kilonova* because it is 1000 times brighter than a nova explosion. This temporal behavior is consistent with energy injection due to heavy element decay, formed by rapid neutron capture (the r-process) in the actinide region. An estimated 1.6×10^4 times the mass of the Earth was calculated to have been produced in this synthesis of heavy elements, including about 10 Earth masses of gold and platinum. This strongly supports the hypothesis that NS–NS mergers are the preferred place for the formation of the heavy elements in the Universe, with $A \sim 200$.

Some final considerations are in order to emphasize the importance of this event. The host galaxy NGC 4993 lies at a measured distance of 40 Mpc as indicated by its redshift of 0.0099. This therefore proved that any event occurring in the Virgo galaxy cluster, residing at half this distance, will be detectable in the future. Since events of this type should also occur at greater distances, it will be possible to give an independent determination of the Hubble constant H_0 with an uncertainty less than 2% within 5–10 years, since these mergers are "standard sirens" that can be calibrated. Other tests of gravitation are also possible, including those that aim to detect the effects of alternative theories of gravitation in the waves. For example, from the relative delay between the gravitational signal and the electromagnetic burst, it was inferred that gravitational effects and photons travel with the same velocity c, with a maximal admissible difference of order 10^{-15}. Finally, the study of the merger itself can bring new information about this and future events, since the observation of a non-zero signal almost 2 s after the event (Fig. 10.14) showed that the black hole did not form immediately. The existence of an intermediate transient state, possibly a supermassive neutron star held in place by high rotation and viscosity, brings an opportunity to study the behavior of matter in this extreme regime. In short, we are witnesses of a revolutionary epoch in Astrophysics with very encouraging prospects in this area in the near future.

References

1. A. Pais, *Subtle Is the Lord: The Science and the Life of Albert Einstein* (Oxford University Press, Oxford, 2005)
2. R. Matzner, *Introduction to Gravitational Waves* (Springer, Dordrecht, Netherlands, 2010)
3. K. Thorne, Gravitational-wave research: Current status and future prospects. Rev. Mod. Phys. **52**, 285 (1980)
4. J.M. Weisberg and J.H. Taylor, The relativistic binary pulsar B1913+16: Thirty years of observations and analysis, in *Binary Radio Pulsars*, ASP Conference Series 328, eds. F.A. Rasio and I.H. Stairs (2005), p. 25
5. J.A. Batlle, R. Lopez, Revisiting the border between Newtonian mechanics and General Relativity: The periastron advance. Contrib. Sci. **10**, 65–72 (2014)
6. M. Burgay et al., An increased estimate of the merger rate of double neutron stars from observations of a highly relativistic system. Nature **426**, 531 (2003)
7. J. Weber, Evidence for discovery of gravitational radiation. Phys. Rev. Lett. **22**, 1320 (1969)
8. P.R. Saulson, Interferometric gravitational wave detectors. Int. J. Mod. Phys. D **27**, 1840001 (2018)
9. B.P. Abbott at el., Prospects for observing and localizing gravitational-wave transients with Advanced LIGO and Advanced Virgo. Living Rev. Relati. **19**, 1 (2016)
10. A. Blaut, Parameter estimation accuracies of Galactic binaries with ELISA. Astropart. Phys. **101**, 17 (2018)
11. https://blogs.cardiff.ac.uk/physicsoutreach/gravitational-physics-tutorial/evolved-laser-interferometer-space-antenna-elisa/
12. C.J. Moore, R.H. Cole, C.P.L. Berry, Gravitational-wave sensitivity curves. Class. Quant. Grav. **32**, 015014 (2015)
13. B.P. Abbott et al., Observation of gravitational waves from a binary black hole merger. Phys. Rev. Lett. **116**, 061102 (2016)
14. B.P. Abbott et al., GW170817: Observation of gravitational waves from a binary neutron star inspiral. Phys. Rev. Lett. **119**, 161101 (2017)

15. J.E. Horvath, The binaries of the NS–NS merging events, in *Proceedings of the Xiamen-CUSTIPEN Workshop on the EOS of Dense Neutron-Rich Matter in the Era of Gravitational Wave Astronomy*, AIP Conf. Proc.2127 (2019), p. 020015
16. R.D. Ferdman et al., Asymmetric mass ratios for bright double neutron-star mergers. Nature **583**, 211 (2020)
17. B.P. Abbott et al., Gravitational waves and gamma-rays from a binary neutron star merger: GW170817 and GRB 170817A. Astrophys. J. Lett. **848**, L13 (2017)
18. S. Covino et al., The unpolarized macronova associated with the gravitational wave event GW 170817. Nature Astron. **1**, 791 (2017)

Chapter 11
Gamma-Ray Bursts

11.1 The Problem of Gamma-Ray Bursts: The Most Distant Objects in the Universe?

As a result of the mutual distrust between the United States and the Soviet Union during the Cold War years, the former began launching a series of satellites carrying instruments to detect clandestine nuclear tests made by their socialist rival. This series, called the *Vela satellites*, operated for several years and managed to detect several short-lived bursts of gamma-rays. But the surprise was that, instead of originating on Earth, they appeared to be arriving from space, and they did not have a solar origin. Theory and observations of GRBs are summarized in this Chapter. As they were part of a secret military project, these data were only made known to the public in 1973, when the project was declassified. Klebesadel, Strong, and Olson [1] announced this discovery and discussed the origin of these bursts in the Astrophysical Journal Letters (Fig. 11.1).

The accumulation of data showed that the arrival directions of the photons and the light curves were unpredictable, with regard to both the duration of the event and its variability (Fig. 11.2). The spectra, however, showed a certain regularity, a fact interpreted as evidence for a generic emission mechanism.

An in-depth study collecting together thousands of events was necessary to answer these questions and to identify the origin of the bursts. The *Compton Observatory* mission was launched in 1991 and continued until 2000 with four complementary on-board instruments: OSSE (a directional scintillator, capable of measuring between 50 keV and 10 MeV), COMPTEL (a two-layer Compton telescope, somewhat similar to an optical camera with sensitivity between 750 keV and 30 MeV), EGRET (a spark chamber to detect the highest energies, between 30 MeV and 30 GeV), and BATSE (Burst and Transient Spectrometer Experiment, an arrangement of 8 modules capable of locating a burst and measuring its spectrum between 20 keV and 8 MeV). All the instruments performed very well, and achieved sensitivities more than sufficient to measure weak bursts and determine the expected flatness of the distribution coinciding with the galactic plane. But this was not what happened: to

© The Author(s), under exclusive license to Springer Nature Switzerland AG 2022
J. E. Horvath, *High-Energy Astrophysics*, Undergraduate Lecture Notes in Physics,
https://doi.org/10.1007/978-3-030-92159-0_11

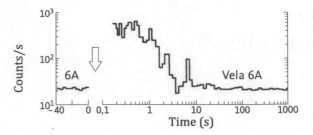

Fig. 11.1 One of the gamma-ray bursts detected by the Vela 6A satellite and revealed in 1973 [1]. The background count went from around 20 to 600 photons/s and decreased again in less than 10 s, as indicated by the *arrow*. © AAS. Reproduced with permission

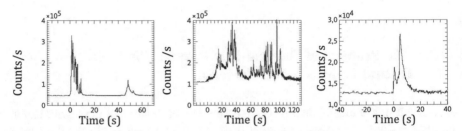

Fig. 11.2 Light curves of bursts with varying complexity and duration, compiled from data in [2]. The presence of very short time scales ($\tau_{min} \sim 0.1$–0.01 ms) in many events suggested that neutron stars might be candidates for producing the bursts: as we saw in Chap. 3, this shorter time implies an emitting region with a size $R \leq c\tau_{min} = 300$ km. The high-energy processes around neutron stars seemed a likely source. The measured fluences indicated released energies of up to 10^{42} erg in gamma rays if the emission was assumed to be isotropic (the fluence is the flux integrated over the duration of the event). But in the data it was not possible to distinguish any spatial bias in the burst distribution, as would be indicated for the distribution of neutron stars associated with the galactic disk—or a more extended region, since we saw that high proper motions are measured for many of these objects, suggesting that they could partially populate the halo in a few million years

the surprise of the whole Astrophysics community, the spatial distribution of the bursts never showed any spatial anisotropy, and the full catalog published by the collaboration [2] is perfectly consistent with an *isotropic* distribution, without any flattening/anisotropy at all (Fig. 11.3).

When the correlation of duration with hardness (Fig. 11.4) became clear, it was natural to assume that the two classes were originated by different phenomena, since hardness is a reflection of the way energy is released. However, another key element, the spectra, showed no clear difference. This was interpreted as due to a quasi-universality of the explosive phenomenon which depended weakly on the specific source. In fact, in Fig. 11.5 we can see a typical spectrum and the so-called *Band parametrization* (see [3] and references therein). The latter is not based on any theory, but well reproduces the vast majority of spectra of short and long bursts. Thus, any successful theoretical model should produce a spectrum compatible with the Band parametrization.

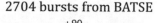

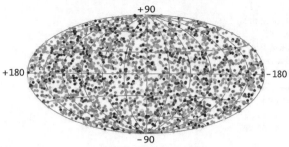

Fig. 11.3 Data compiled from the final BATSE catalog [2] showing a statistically isotropic spatial distribution, a fact that contradicts local neutron stars as sources. Starting with the announcement of this result in 1993, and for each extension of the catalog with new bursts, it became increasingly difficult to imagine a galactic origin, even for objects located in an extended halo—note that, if so, there should be an excess of events in the direction of the Andromeda galaxy, but this was never observed. Of course, there is a distribution of sources that would be "naturally" isotropic around us: the cosmological one. But this in turn would require a gigantic luminosity, greater than 10^{50} erg of gamma rays alone, and it was not obvious how to reach this scale, nor how to leak the gammas from the source without their being degraded to lower frequencies (see below). The other important discovery was that the duration of the bursts separated them into (at least) two classes: the "short bursts" of up to about 2 s, and the "long bursts", with durations of 10 s or more (see next figure). Credit: Michael Briggs, NASA [4]

The Band parametrization reads

$$N(E) = AE^\alpha \exp -(E/E_0) \quad \text{(low energy)}, \tag{11.1}$$

$$N(E) = BE^\beta \quad \text{(high energy)}, \tag{11.2}$$

where $\alpha = -1$, $\beta = [-2, -3]$ and $E_0 \approx 150$ keV. The value of the exponent β is the most variable, and can reach -3 for some cases where the spectrum falls off very quickly with the energy (Fig. 11.5).

All the ingredients were now available for the elaboration of theoretical models that would explain the nature of the bursts and reproduce the observations of light curves and spectra. Before confirmation, and remembering the belief that the bursts were produced by nearby neutron stars, more than 100 theoretical models were published, including comets falling on neutron stars, magnetic instabilities, and many other ideas [6]. But having only gamma-ray data and no other information, it was impossible to make any real progress in this matter. Accurate location of the bursts was essential in order to confirm the distance scale, and with it the energy. Gamma-ray data take time to process and the error circle was typically at least 1°, insufficient for such quick follow-up research by optical telescopes.

Despite the delay of almost a decade, the launch of the Italian–Dutch BeppoSAX mission, which carried a wide field camera operating between 2 and 30 keV, was fundamental to establish the location of the bursts. In 1997 BeppoSAX detected a

Fig. 11.4 Histogram of
burst durations in the BATSE
sample, showing a bimodal
distribution (*upper*) [5].
These two types of bursts
also have more or less
energetic (or "hard")
photons, so the short bursts
are harder than the long
ones—see, for example,
[3, 8] and references therein

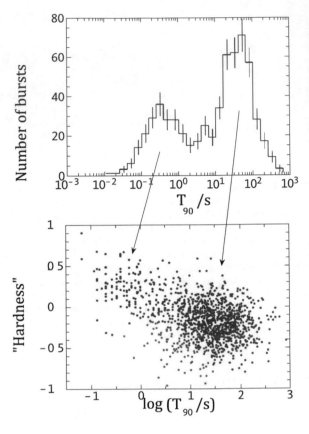

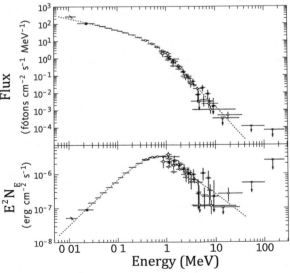

Fig. 11.5 Typical GRB
spectrum [7]. As the flux
drops enormously with the
energy (*top panel*), it is usual
to multiply the number of
photons N by E^2 to
highlight the value where the
maximum is reached (*bottom
panel*). The Band
parametrization is the *dotted
line*

Fig. 11.6 *Upper*: Images of GRB 970508 obtained by BeppoSAX at 5 day intervals. Credit: BeppoSAX. *Lower*: Optical transient (known as the *afterglow*) observed by the Hubble telescope while the intensity was still rising. Credit: STScI/NASA–HST/STIS

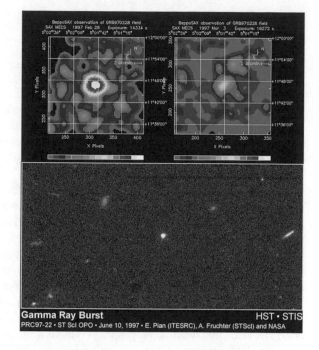

gamma burst (GRB 970508) and located it with an accuracy of a few arcmin [9] thanks to the X-ray detector's ability to collimate the incoming radiation and refine its spatial origin (Chap. 3). Some 5 hr after the burst, an optical transient was detected of magnitude around 20 and which rapidly faded. The host galaxy was thus identified and the extra-galactic origin of the bursts finally demonstrated (Fig. 11.6).

It was thus clear that we were dealing with a gigantic energy scale, and that nearby neutron star models were inadequate by several orders of magnitude. The magnitude of the inconsistency can be quantified by stating the problem of compactness. The detected spectra were clearly non-thermal (Band-type). Thus, the *transparency* to gamma photons had to be $\hat{\tau} < 1$, otherwise something like a black body would be observed. But from the observed variability, it was also clear that the sources were highly compact, and if they were extragalactic, energies above 10^{50} erg in such compact dimensions would imply opacities $\hat{\tau} \gg 1$. The spectra and the compactness were two incompatible things in a first analysis.

11.2 Models of the Bursts

Although these problems seem insoluble at first sight, some studies immediately found a "kinematic" way out of the problem that is generally accepted until today, namely that the ejected material that emits the gammas moves with ultra-relativistic speed. If this is the case, the material is transparent in the proper reference frame, since the optical depth $\hat{\tau}$ is reduced by kinematics to

$$\tau_{\text{eff}} = \hat{\tau} \left(\frac{\epsilon_{\max}}{m_e c^2} \right)^{\alpha-1} \gamma^{-2(\alpha-1)} , \tag{11.3}$$

where $\epsilon_{\max}$ is the maximum photon energy and the α index varies for each burst, but generally oscillates between 2 and 3. Equation (11.3) shows that τ_{eff} can be made less than unity if the Lorentz factor is $\gamma > 100$, which compensates for the high value of $\hat{\tau}$. For comparison, the AGN jets discussed in Chap. 8 have Lorentz factors 5–10 [10].

Kinematically an ultra-relativistic flow is observed in the laboratory frame with a kinematic beaming factor dependent on γ (already shown in Fig. 2.13), sometimes also called the *emission cone*. This should not be confused with a geometric collimation (the latter may also happen in the proper frame), but has an important effect on the evaluation of the emitted energy: if the emission cone is small, since its opening angle is $\theta \sim \gamma^{-1}$, the actual emitted energy is much less than its isotropically estimated value. Frail et al. [11] found that all GRBs have a unique energy of about 10^{50} erg when they analyzed the BATSE sample, and concluded that the emission cone angles are less than 5°. This is why we often talk about *equivalent isotropic energy*: a value that would be appropriate if there were no beaming and the emission were genuinely isotropic. These equivalent isotropic energies can reach 10^{54} erg. But there is yet another consequence for "alleviating" the energy problem: since most of the bursts do not emit in our direction, their actual number must be of the order of 100 times what is observed (!).

To explain these features within a theoretical model, the so-called *fireball model* was put together, with contributions from several scientists over a decade. Essentially, the fireball model requires an event that injects almost pure "radiation" jets (i.e., with little baryonic content, to allow a large Lorentz factor, as required to lower the opacity) in an episodic manner. Episodic ejection ensures that each ejected bubble has its own Lorentz factor, and when the fastest ones reach the slowest ones ahead, internal shocks are produced. These shocks produce the observed gamma-rays, since the opacity is reduced according to (11.3). When this jet collides with the ISM, it decelerates, and when the non-relativistic regime is reached, it opens geometrically letting out softer radiation, from X-rays to radio, which constitutes the so-called afterglow of the bursts. This scenario is shown in Fig. 11.7.

The most convincing proof of the kinematic effect of the opening leading to the afterglow is the so-called *spectral break* observed in many events. Figure 11.8 illustrates this observation and its interpretation. We must point out, however, that some bursts do *not* present any spectral break. It is not clear whether these are truly isotropic, but if so, they should contain energies up to 1000 times greater than the average. A wide variety of light curve behaviors is still present in the sample, possibly because the variation of the basic fireball scenario requires this.

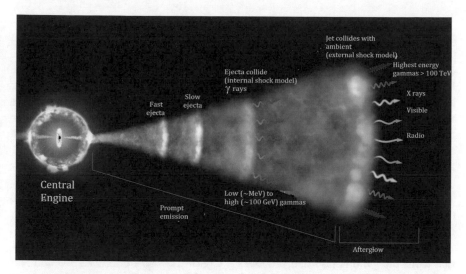

Fig. 11.7 Fireball model. Two types of event produce a "central engine", namely, a merger (short GRB) and a collapsar (long GRB). When the ultra-relativistic jets are ejected, the internal and external collisions produce the gamma-ray burst and the afterglow, respectively. Credit: NASA/Goddard Space Flight Center/ICRAR

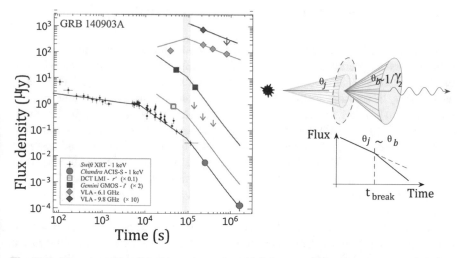

Fig. 11.8 The spectral break is the moment when the light curve changes slope in time. *Left*: The break in the GRB 140903A event, which is achromatic, contrary to most of the cases [12]. *Right*: The break time is assigned to the moment when the jet becomes non-relativistic and suddenly opens wide, since $\theta \sim \gamma^{-1}$. This explains the more isotropic character of the afterglows when compared with the gamma emission

Finally, there is the question of the "event" (engine) itself, which needs to produce an ultra-relativistic flow with "short" or "long" duration as observed. After much discussion of this issue, the basic consensus is that there are two prominent candidates: neutron star fusion and the collapse of a very massive star to form a black hole. In both cases, the intermediate configuration is practically the same, namely, a black hole with a transient accretion disk, as shown in Fig. 11.7. In the case of a neutron star merger, numerical simulations show that a duration up to 2 s is reasonable for this event (although the efficiency of the energy conversion to gamma-rays obtained in the jet remains somewhat uncertain). In the case of collapse (also called a *hypernova*), it seems that each time a black hole is formed there will be an associated "long" burst, produced by the injection of a jet perpendicular to the disk plane. These models need the jet to "punch" the envelope of the collapsing star, and it is not entirely clear how this happens. But the important thing is that these two events are consistent with the temporal bimodality of the bursts, produce an analogous intermediate state (black hole plus disk, but with different duration), and launch a fireball that emits the gamma-rays and produces afterglows in most cases. As we will see below, certain observations suggest that this identification is feasible and that at least these two basic models can explain the events, although new scenarios may emerge to give more specific features in some subset of the events [13].

11.3 Recent Observations and Models of GRBs

Like any theoretical construction, the fireball model produced by the merging of neutron stars or hypernova events requires factual confirmation, and it is this evidence that we will discuss here. As we saw in the previous chapter, the event GW 170817 proved that, in addition to the gravitational signal, a short-lived gamma burst is produced [14]. Other previously known events already pointed in this direction: GRB 130603B was a short burst with the emergence of infrared radiation one week after the gamma event. This was interpreted as evidence that the ejected material had produced lanthanide nuclei of high opacity, a fact that corresponds very well to expectations when two neutron stars merge, as simulated theoretically. Thus, we can speak of a validation of the merger model, and future events of this type should confirm this idea.

On the other hand, there is also evidence that the hypernova model works in practice for "long" bursts. This conclusion stems from the existence of events where a long burst is first detected, while the spectrum is observed to "transform" after several hours into a corresponding supernova (of type Ib or Ic, i.e., a collapse where the progenitor has lost its envelope, discussed in Chap. 5). The first known case was that of GRB 980425, followed by the associated SN 1998bw, which had ejection speeds of about 30 000 km/s. Today, some forty cases of this type are known, and the conclusion is that the "hypernova" model really works for the production of a burst, although it is also clear that a subset of the "long" bursts does not have any associated supernova (and this would be the case if they could also be produced

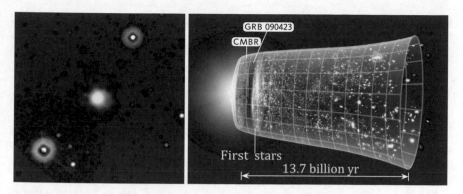

Fig. 11.9 GRB 090423 (*left*) and its location in the expansion diagram of the Universe (*right*), preceding most star formation (inside the dark era, before the massive formation of stars and galaxies, as indicated). Credit: Swift X-ray Telescope NASA and NASA/WMAP

by another explosive event, to be confirmed). The corollary is that the birth of a stellar-mass black hole is "announced" by a "long" gamma-ray burst.

Precisely, this last conclusion is the one that allows us to infer that gamma-ray bursts can be an important tool to study the structure formation stages of the Universe. This statement is deduced from the fact that there are events where it has been possible to locate the source at extreme distances. The most interesting case is that of GRB 090423 (Fig. 11.9). In the detected afterglow of this extremely intense event, neutral hydrogen absorption lines have been detected with various redshifts. For extragalactic astrophysicists, this is a clear sign of cosmological distance, since this absorption due to neutral hydrogen clouds along the line of sight (the so-called Lyman-α forest) is observed in the remote quasar spectra. The measured redshift was $z = 8.2$, that is, the burst happened before any significant formation of stars in bulk, and this burst was the most distant object ever detected at the time of its detection (today we know of a few galaxies that were formed even earlier).

This type of detection can be thought of as evidence in favor of explosive events, possibly from stars that formed very early on with huge masses—called *Population III* by astronomers—and would therefore have evolved very rapidly, giving rise to the formation of high-mass black holes, perhaps similar to those detected in GW 190521 [15] (presented in the last Chapter). If so, and according to the rate measured by BATSE, INTEGRAL, and other instruments, 1–2 black hole births are witnessed every day in the observable Universe, and something like 100 times this figure actually occur if the beaming ideas are correct.

Finally, there are controversies about using the bursts as "standard candles" in order to measure the Hubble constant. If possible, this would be highly desirable, since the bursts are very bright and can be measured even when they occur at huge distances (such as GRB 090423). But the intrinsic luminosity of the bursts seems to vary a lot, that is, they are not "standard candles" *prima facie*, and it is not clear to

what extent they can be used for these purposes, although this has not yet prevented
work from being done along these lines.

11.4 Fast Radio Bursts: A Related Phenomenon?

One lesson from the history of Astronomy is that novel observational techniques
and strategies have great potential for discovery, and often lead to completely new
and unexpected phenomena. This was certainly the case for GRBs and many other
objects, and also for the more recent case of *fast radio bursts (FRB)*.

The first confirmed report of one of such an event was a short, isolated pulse in
radio frequencies (see Fig. 11.10 upper panel), of millisecond duration, observed by
Lorimer [16], although there is a possibility that other events may have been seen
earlier by other groups. After collecting more events, it turns out that the typical
emission frequencies are around 1 GHz, and possibly down to a few hundred MHz.
On the other hand, the emission bandwidth is important to estimate the distance to
the source, as we shall see below. Because of this short duration, their occurrence can
be very high and go unnoticed, as already pointed out in [16]. Careful examination
thus revealed a hidden phenomenon.

Fig. 11.10 *Upper*: A typical
FRB spike in radio
frequencies [18, Fig. 2].
Lower: Location of a set of
FRBs in galactic coordinates
[17]

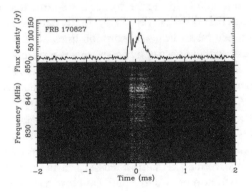

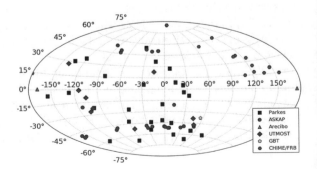

A very precise measurement of the arrival time is important to characterize the distances as follows. If ν_{inf} and ν_{sup} are the lowest and highest frequencies of the event, the calculated delay between them is [17]

$$\Delta t = \frac{e^2}{2\pi m_e c} \left(\frac{1}{\nu_{\text{inf}}^2} - \frac{1}{\nu_{\text{sup}}^2} \right) \times DM , \qquad (11.4)$$

where m_e is the electron mass and DM is the *dispersion measure* given by

$$DM = \int_0^d n_e(l) dl , \qquad (11.5)$$

i.e., the integrated electron density along the line of sight up to the source at a distance d. The pulse is dispersed by the electron clouds and the effect on the delay of the limiting frequencies can be measured directly, allowing a determination of the DM. With these expressions, it became clear that most of the events are extragalactic. In some cases, $DM \geq 2000$ has been inferred, corresponding to cosmological distances. Therefore, the energy scale and the spatial distribution should correspond to a cosmological population (Fig. 11.10).

It is clear from this evidence that the FRBs are extremely luminous, although the energy released is not that huge because of their short duration. A comparison with other known transitory and stationary sources may be attempted, in terms of a spectral luminosity vs. duration diagram (Fig. 11.11), in which the vertical axis is just the luminosity divided by the bandwidth of the emission. FRBs occupy the upper left region, with a spectral luminosity comparable to the AGNs of Chap. 8, and typically higher than GRBs and supernovae. Because of this feature and the associated brightness temperature, it is agreed that they are an outstanding example of *coherent* emission, separated from the incoherent sources filling the grey sector in Fig. 11.11. This will be important for any attempt to find viable physical models for these bursts.

Important news for the field has been presented recently, with the detection of repeating sources (of unknown origin) and also the identification of a galactic magnetar (Chap. 6) as the source of nearby FRBs [19]. Figure 11.12 is a simple diagram to illustrate the recurrent behavior of FRBs, with the magnetar-associated events (ST 200428A marked in Fig. 11.11) included in the "repeaters" set. The direction was coincident with the magnetar SGR 1935+2154 and the small DM indicated a nearby galactic origin.

Physical models of FRBs are presently "in the works". While it is tempting to associate the extragalactic/cosmological sources with some kind of catastrophe, it must be taken into account that no high-energy emission coincident with the radio pulse has yet been detected. Therefore, it is unlikely that FRBs are the "tail" of some outburst like the giant flare of a magnetar in Fig. 6.22. However, there are some promising ideas, such as the Falcke–Rezzolla scenario [20] in which the col-

Fig. 11.11 Spectral luminosity vs. duration on a log–log scale. Several known sources and FRBs are located to give an idea of their energetic behavior [19]

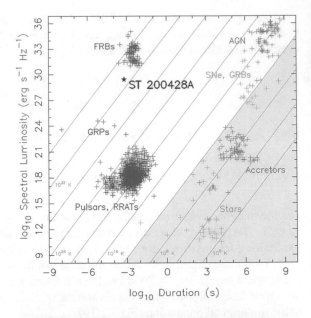

Fig. 11.12 Simplest classification of the temporal behavior of FRBs

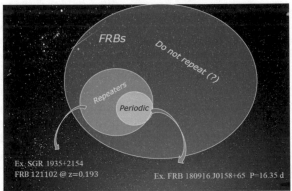

lapse of a briefly living supramassive neutron star produces a radio burst when the magnetosphere "snaps" before disappearing behind the event horizon.

On the other hand, the association of a handful of bursts with a nearby magnetar revived models in which a quake of the crust is the cause of a sudden energy release. Shaking of field lines can launch Alfvén wave propagation, which eventually radiate high up in the magnetosphere due to the curvature mechanism (similar to the synchrotron expressions of Chap. 2 [21]). The detection of X-rays almost simultaneous with the FRB pulses [22–24] holds an important clue for the generation mechanism of the bursts, one that remains to be understood. In this sense, the FRBs are likely related to the GRBs: the majority are cosmological, compact objects are involved in their generation, and their historical development parallels the latter. They will remain a hot topic for years to come.

References

1. R.W. Klebesadel, I.B. Strong, R.A. Olson, Observations of gamma-ray bursts of cosmic origin. Astrophys. J. **182**, L85 (1973). https://doi.org/10.1086/181225
2. W.S. Paciesas et al., The fourth BATSE gamma-ray burst catalog (revised). Astrophys. J. Supp. **122**, 465 (1999)
3. C. Guidorzi, http://www.fe.infn.it/~guidorzi/doktorthese/node5.html (2003)
4. https://gammaray.nsstc.nasa.gov/batse/grb/skymap/
5. S.D. Barthelmy, *Swift-BAT results on the prompt emission of short bursts*, Phil. Trans. Roy. Soc. A **365**, 1281–1291 (9 February 2007)
6. R.J. Nemiroff, *A century of gamma ray burst models*, Comm. Astrophys. **17**, 189 (1994). Available at arXiv:astro-ph/9402012
7. R. Willingale, P. Mészáros, Gamma-ray bursts and fast transients. Space Sci. Rev. **207**, 63–86 (2017). https://doi.org/10.1007/s11214-017-0366-4
8. A. Shahmoradi, R.J. Nemiroff, Short versus long gamma-ray bursts: a comprehensive study of energetics and prompt gamma-ray correlations. MNRAS **451**, 126 (2015)
9. L. Piro et al., Evidence for a late-time outburst of the X-ray afterglow of GB970508 from BeppoSAX. Astron. Astrophys. **331**, L41 (1998)
10. P. Mészáros, The fireball model of gamma-ray bursts. Prog. Theor. Phys. Supp. **143**, 33 (2001)
11. D. Frail et al., Beaming in gamma-ray bursts: Evidence for a standard energy reservoir. Astrophys. J. Lett. **562**, L55 (2001)
12. E. Troja et al., An achromatic break in the afterglow of the short GRB 140903A: evidence for a narrow jet. Astrophys. J. **827**, 102 (2016)
13. I. Horváth, A further study of the BATSE gamma-ray burst duration distribution. Astron. Astrophys. **392**, 791 (2002)
14. E. Waxman et al., Constraints on the ejecta of the GW170817 neutron star merger from its electromagnetic emission. MNRAS **481**, 3423 (2018)
15. R. Abbott et al., Properties and astrophysical implications of the $150 M_\odot$ binary black hole merger GW190521. Astrophys. J. Lett. **900**, L13 (2020)
16. D. Lorimer et al., A bright millisecond radio burst of extragalactic origin. Science **318**, 777 (2007)
17. E. Petroff, J.W.T. Hessels, D. Lorimer, Fast radio bursts. Astron. Astrophys. Rev. **27**, 4 (2019)
18. W. Farah et al., FRB microstructure revealed by the real-time detection of FRB170827. MNRAS **478**, 1209 (2018)
19. C.D. Bochenek et al., A fast radio burst associated with a galactic magnetar. Nature **587**, 59 (2020)
20. H. Falcke, L. Rezzolla, Fast radio bursts: the last sign of supramassive neutron stars. Astron. Astrophys. **562**, A137 (2014)
21. M. Longair, *High-Energy Astrophysics* (Cambridge University Press, Cambridge, 2011)
22. S. Mereghetti et al., INTEGRAL discovery of a burst with associated radio emission from the magnetar SGR 1935+2154. Astrophys. J. Lett. **898**, L29 (2020)
23. C.K. Li et al., HXMT identification of a non-thermal X-ray burst from SGR J1935+2154 and with FRB 200428. Nature Astronomy **5**, 378 (2021)
24. M. Tavani et al., An X-ray burst from a magnetar enlightening the mechanism of fast radio bursts. Nature Astronomy **5**, 401 (2021)

Chapter 12
Cosmic Rays

12.1 Messengers from the Greatest Accelerators in the Universe: Cosmic Rays

The history of the discovery of cosmic rays is fascinating in itself and corresponds to a pioneering phase of Astrophysics in the early 20th century, when Relativity and Quantum Mechanics changed our perspective of the physical world. In this context the Austrian physicist Victor Hess was studying the radioactivity of elements (then only recently discovered) and related problems, and decided to attempt direct measurement of the degree of ionization in the high atmosphere. This problem was intriguing because there was evidence in favor of an increase in ionization with height. Hess refined an electroscope to measure ionization and took it in person over 3 km up with the help of balloons (Fig. 12.1) between 1911 and 1912. His data showed that this result was correct, and that it depended little on whether the measurement was made during the day or at night. The data obtained during a solar eclipse were particularly important for attempts to characterize the phenomenon. The origin and propagation of cosmic rays are addressed here, in particular, the so called Ultra-High energy range with all the associated puzzles and questions.

When this result was confirmed, it became evident that some source of ionization was coming from outer space. Although the energy required to explain the measured effect was low, the possibility of high-energy ionizing particles, even outside the accessible range of the Hess spectroscope, remained open. The bewilderment of physicists at the time is evident if we consider the name they chose for them: cosmic rays. Only later was it clarified that these "rays" were actually electrons, protons, and nuclei, which justifies having spoken of "particles" in the previous sentences.

With the passing of time and considerable effort by many groups in various places on the planet, the general form of the spectrum of these cosmic rays was eventually detected and established. This spectrum is shown in Fig. 12.2. There are several important features in it, some of which have been discussed and studied for over a century [1].

© The Author(s), under exclusive license to Springer Nature Switzerland AG 2022
J. E. Horvath, *High-Energy Astrophysics*, Undergraduate Lecture Notes in Physics,
https://doi.org/10.1007/978-3-030-92159-0_12

Fig. 12.1 Victor Hess and collaborators in one of the balloon flights that led to the confirmation that ionization increases with height in the atmosphere. Credit: V.F. Hess Society

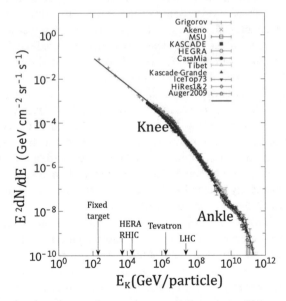

Fig. 12.2 Spectrum of cosmic rays, from the lowest energies up to 10^{20} eV, the highest energies observed. The experiments that contributed to this diagram are identified with their error bars. The number of particles per energy range dN/dE on the vertical axis is usually multiplied by E^2 to improve visualization. Around 3×10^{15} eV there is a change in the slope known as the "knee", while at 10^{18} eV a second change in the slope became known as the "ankle". Above this last energy the flux is very small, around one particle/km^{2}yr, and gigantic experiments are needed to accumulate enough statistics. The center of mass energies of the Fermilab, CERN, and other experiments are indicated by *arrows*, orders of magnitude lower than those produced by accelerators in nature. Figure adapted from [2]. For an update, see [3]

One of the striking features of cosmic rays in the lower energies is that the flux is modulated by the solar wind, which can sweep away particles of cosmic origin, thereby "protecting" Earth's biosphere where its intensity is greater. We could say that Earth's magnetic field is our protector. Without it life on the planet's surface would be barely possible. Bombardment by the particles discovered by Hess would be lethal to living beings if they struck directly. On another note, the highest energy regime is a real challenge. The detection of primaries with energies above 10^{20} eV is equivalent to putting the energy of a bullet into an elementary particle (!). So nature has some mechanism for accelerating even up to these extreme energies.

12.1.1 Origin, Propagation, and Acceleration

The natural question is: where do these particles come from? An examination of the spectrum shows a change in the spectral index α at the knee and ankle, which suggests that the origin is different above and below the energies where the changes occur. This idea has important support from the point of view of propagation, as we will see below.

Let us consider the relativistic momentum $\mathbf{p} = \gamma m \mathbf{v}$ of a charged particle. As we saw in Chap. 2, the *gyroradius* or Larmor radius is

$$r_L = \frac{p_\perp}{|q|B} = \frac{\gamma m v}{ZeB} , \qquad (12.1)$$

where q is the electric charge and $p_\perp$ the transverse component of the momentum. Numerically, for the case of a proton (more than 90% of cosmic rays are protons), we have

$$r_L = \frac{(pc/\text{MeV})}{300(B/\text{G})} \text{ cm} . \qquad (12.2)$$

Using (12.2), we can calculate the maximum energies at which the Larmor radius becomes equal to the scale of interest L. The particles can rotate in confined paths if $r_L < L$. Starting with the heliosphere with $L = 100$ AU and where $B \sim 10 \, \mu$G, we have $E_{\text{max}} \sim 4 \times 10^{12}$ eV. This shows that the solar modulation happens to the left of Fig. 12.2, at low energies, as claimed.

For the typical scale of the ISM, we have instead $L = 100$ pc and $B \sim 5 \, \mu$G, and therefore $E_{\text{max}} \sim 4 \times 10^{17}$ eV. Particles with these energies do not suffer substantial deflection in the galaxy. At intermediate energies, we know that the "knee" occurs at around 10^{15} eV, and below around 10^{14} eV observations show that the cosmic rays are totally isotropic. The interpretation is that they are confined within the galaxy. Since above 10^{15} eV, confinement is not possible, it seems likely that above these primary energies the particles come from outside the galaxy. The observed change in the α index from 2.7 to 3 at the "knee" would be a consequence of this. It is also possible that at 10^{14} eV it is the very mechanism of acceleration that changes, and that the extragalactic component corresponds to cosmic rays arriving with $E > 10^{18}$ eV.

We can quantify this change in the spectral slope a little more closely if we assume that the primaries are all protons and that they diffuse from place to place in the galaxy. In a diffusion process of this type, there is a diffusion time—similar to the one defined for the photons in (4.25)—given by the expression $t_{esc} = R^2/D$, where D is the diffusion coefficient, which depends on the energy of the particles. A simple but accurate expression for this quantity is $D \approx 0.1c(E/1\,\text{GeV})^{-1/2}$. Assuming a distance scale $R \ll 1$ kpc, suitable for keeping protons within the galactic disc, leads immediately to

$$t_{esc} = 4 \times 10^7 \left(\frac{E}{1\,\text{GeV}} \right)^{-1/2} \text{yr} . \tag{12.3}$$

Thus, if we consider that the sources inject a spectrum $dN/d\ln E|_{\text{inj}}$, the observed spectrum in the diffusive regime will be different, given approximately by

$$\left. \frac{dN}{d\ln E} \right|_{\text{obs}} \approx \left. \frac{d\dot{N}}{d\ln E} \right|_{\text{inj}} t_{esc} , \tag{12.4}$$

which results in a prediction that the spectrum injected (accelerated) by sources has $d\dot{N}/d\ln E|_{\text{inj}} \propto E^{-2.2}$ for protons of energy around 1 GeV. We will soon see that the acceleration by the Fermi mechanism gives a very similar value for this exponent.

A more detailed comparison between the abundances measured in the solar vicinity and those of the cosmic rays can be appreciated in Fig. 12.3. We see that some elements in the cosmic rays (open circles, full line) are much more abundant than in the Solar System (filled circles, dashed line). In particular, lithium, beryllium, boron, scandium, and vanadium have abundances several orders of magnitude greater in cosmic rays. The consensual explanation for this difference is that there is a production process, known as *spallation*, between the primary protons and very abundant nuclei (such as carbon and oxygen) that fragment the latter and produce the overabundant elements. Thus, lithium, beryllium, and boron are carbon and oxygen "chips", and the heavier ones like scandium and vanadium are "chips" of iron, which is also abundant in the solar neighborhood.

Fig. 12.3 Abundances measured in cosmic rays (*open circles*) vs. Solar System values (*filled circles*)

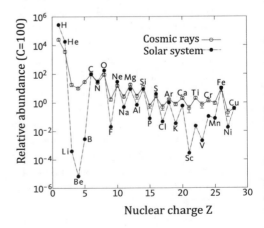

Consider now the problem of the acceleration of primaries (see, for example, [4]). Certainly, the simplest way to accelerate a charged particle is to submit it to electromagnetic forces. We know in a very general way that the equation of motion of a charged particle in an electromagnetic field is (2.22), viz., $d(\gamma m\mathbf{v})/dt = q\mathbf{E} + q(\mathbf{v} \times \mathbf{B})$. The first term on the right can lead to a *direct acceleration* as long as $\langle \mathbf{E} \rangle \neq 0$, but this is difficult in Astrophysics, since the process of separating charges to produce an electric field never lasts long enough to produce a potential because it is quickly neutralized by electrons and positrons from the environment. The exception to this is the process of *magnetic reconnection*, where force lines change their configuration and generate very strong local electric fields. The other way of accelerating particles is to achieve *cumulative gains of energy* by means of random collisions in some medium that has the capacity to support this energy transfer to the particles.

In 1949 Enrico Fermi worked on the problem of energy gains from a particle that collides with low velocity clouds, as happens in the interstellar environment (Fig. 12.4). Fermi showed that after a long time the primary particle which bounces back in the discontinuities shown in Fig. 12.4 gains an average energy of $\langle \Delta E/E \rangle = (8/c)(v/c)^2$. Because the gain is proportional to the small quantity $(v/c)^2$, this result became known as the *second order Fermi mechanism*. Besides the low efficiency, the energy losses were not taken into account. Thus, Fermi considered the possibility of some process where the particle gained energy in *each collision*, and preferably more efficiently.

The idea of this process is that a particle gains energy by going back and forth through a *shock discontinuity*, as illustrated in Fig. 12.5. In a very general way, if $E = \beta E_0$ is the average energy after each passage, and P the probability that the particle remains in the region of acceleration, after k passages, there will be $N = N_0 P^k$ particles with energy $E = E_0 \beta^k$. Therefore, it is possible to eliminate k from these relationships by the manipulation

$$\frac{\ln(N/N_0)}{\ln(E/E_0)} = \frac{\ln P}{\ln \beta} \longrightarrow \frac{N}{N_0} = \left(\frac{E}{E_0} \right)^{\ln P/\ln \beta}, \qquad (12.5)$$

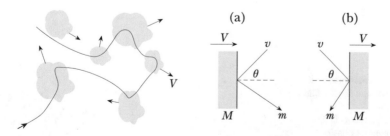

Fig. 12.4 *Left*: A particle with the continuous path shown collides successively with clouds that have random speeds. *Right*: Particles are reflected in the mid-cloud interface and gain energy when the collision happens in situation (**a**) (forward), but lose in situation (**b**) (backward), depending on the relative direction between the particle's velocity and that of the cloud

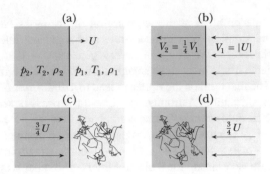

Fig. 12.5 A shock with velocity U seen in (**a**) the observer's system, (**b**) the system moving with the shock, (**c**) the upstream reference system (where matter has not yet been hit by it), and (**d**) the downstream system. The particle goes back and forth, and each time sees the plasma come down on it with velocity $v = 3U/4$

and thus obtaining a theoretical power-law expression:

$$N(E)\mathrm{d}E \propto E^{-1+\ln P/\ln\beta}\mathrm{d}E \ . \tag{12.6}$$

For each process we must calculate the energy gain per cycle to obtain P and β, and thereby determine the shape of the injection spectrum of this process.

In the Fermi process where the particle passes through the shocks, from the downstream reference frame where the plasma approaches with velocity v, the energy of the particle is

$$E_{\mathrm{down}} = \gamma(E + p_x v) \ , \tag{12.7}$$

with p_x the momentum perpendicular to the shock. For non-relativistic situations, it would be enough to put $\gamma = 1$, but for relativistic shocks with $\gamma > 1$, we have $p_x = (E/c)\cos\theta$ (Fig. 12.6). When the particle passes through the shock, the energy changes by

$$\Delta E = pv\cos\theta \ , \tag{12.8}$$

and therefore,

Fig. 12.6 Incidence of a particle on the shock. The momentum $p_x = (E/c)\cos\theta$ is a consequence of basic Mechanics and the relativistic hypothesis $E = pc$

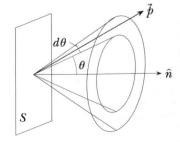

$$\frac{\Delta E}{E} = \frac{v}{c} \cos \theta \ . \tag{12.9}$$

Now, as can be seen in Fig. 12.6, the probability that a particle strikes at an angle between θ and $\theta + d\theta$ is proportional to $\sin \theta \, d\theta$. Its approach speed is $v_x = c \cos \theta$. The probability distribution of the angles is then $p(\theta) = 2 \sin \theta \cos \theta \, d\theta$. Using this we can calculate the angular mean of the gain in (12.9) to take into account a very large set of particles, as we would expect in real acceleration events. This average is

$$\left\langle \frac{\Delta E}{E} \right\rangle = \frac{v}{c} \int_0^{\pi/2} 2 \cos^2 \theta \sin \theta \, d\theta \ , \tag{12.10}$$

interpreted as the average gain of any particle passing through the shock once. As it does not matter whether the particle went forward or backward, this gain is valid for both cases, and a round trip cycle changes the initial energy by

$$\beta \equiv \frac{E}{E_0} = 1 + \frac{4}{3} \frac{v}{c} \ . \tag{12.11}$$

We see that the gain is linear in the speed v, whence it was referred to as the *first order Fermi mechanism*, a much more efficient process than the second order process mentioned above.

We are now in a position to connect these results with the expected power law for this mechanism given in (12.6). For this purpose, we have immediately that

$$\ln P = \ln \left(1 - \frac{U}{c} \right) = -\frac{U}{c} \ , \tag{12.12}$$

$$\ln \beta = \ln \left(1 + \frac{4}{3} \frac{v}{c} \right) \approx \frac{4}{3} \frac{v}{c} = \frac{U}{c} \ , \tag{12.13}$$

which leads to $\ln P / \ln \beta = -1$. Substituting into (12.6), we have

$$N(E)dE \propto E^{-2}dE \ . \tag{12.14}$$

This calculation is still rather heuristic. To obtain the "correct" spectrum we would need to include the losses, the magnetic field of the environment, etc., of the primaries represented by the variable $N(E)$. The complete diffusion equation for this problem would be something like [1]

$$\frac{dN}{dt} = D\nabla^2 N + \frac{\partial}{\partial E}[b(E)N] - \frac{N}{t_{esc}} + Q(E) \ , \tag{12.15}$$

with $D\nabla^2 N$ the diffusive term (previously ignored), $b(E) = -dE/dt$ the energy losses (also ignored here), t_{esc} the escape time of (12.3), and $Q(E)$ the function that describes the sources of the primaries. There are interesting and illustrative solutions of (5.14) in the literature, but they are outside the scope of this introduction (although they delight mathematical Physics enthusiasts).

Analysis of the first order Fermi mechanism suggests that sufficiently strong and extensive shocks (such that the Larmor radius of the particles does not exceed the shock radius, thus ensuring a high number of passages that energize the particles) might be candidates to accelerate the primaries. In addition, shocks should be common to explain the observed energy density of primaries, viz., around 1 eV/cm^3. Supernova remnants meet all the conditions, and the way to check that they are places where particles may be accelerated is very interesting: the acceleration of electrons is easier (although subject to losses by synchrotron radiation, etc.), but accelerated protons would produce gamma-rays by means of the reaction

$$p_{cr} + p_{amb} \rightarrow X + \pi^0 \rightarrow X + \gamma\gamma \,, \tag{12.16}$$

where p_{cr} is an accelerated proton in the remnant and p_{amb} a "fixed" proton standing in the ambient, X any hadronic species in which we have no interest, and the neutral pion π^0 inevitably decays into gammas. Thus, the acceleration signature would be gamma emission at around 100 TeV. Sometimes this expected emission is referred to as the *pion bump*. The study of supernova remnants to identify non-thermal emission that would indicate proton acceleration and the reaction in (12.16) was an exercise that took many years, but recently it was unequivocally confirmed that IC 443, W28, Kepler, and other remnants do indeed have regions where this emission exists (Fig. 12.7) [5]. The maximum confirmed energy is of the order 10^{15} eV [6], compatible with the prediction that the remnants are mainly responsible for the primaries with lower energies than those of the "knee". There is still evidence that the shock that advances through the ISM is spending energy precisely to accelerate particles, that is, the sharing of energy should include these accelerated particles.

The composition aspect of primaries has been mentioned briefly before. We have already stated that almost 90% in the energies lower than the knee are protons, 10%

Fig. 12.7 The supernova remnant observed by Johannes Kepler in 1604. In the south region marked by the *arrow*, observations show emission with a non-thermal spectrum that is compatible with the suggested proton acceleration. Credit: NASA/ESA/JHU/R. Sankrit & W. Blair

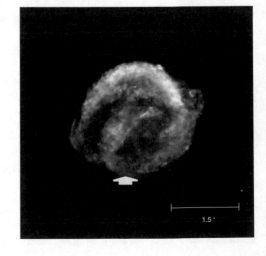

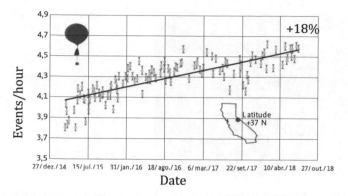

Fig. 12.8 Measurements of cosmic ray events in intermediate energies up to about 30 MeV over a period of more than a year. The increase in the number of events is clear. Credit: H. Takai

are α-particles (helium nuclei), and the rest are decreasing proportions of heavier nuclei. Above the "knee" things are more complicated, since events are much rarer, there are large fluctuations from event to event, and even hadronic interactions are uncertain, especially in the higher energies above the "ankle". We will return to this subject in due time.

Let us finish this discussion with two interesting subjects that still have no definitive answers. The first is the relationship between the cosmic ray primaries and the biosphere. Although sunspots were regularly seen at the solar surface in the past, there was a period until recently (2016) when the Sun was not very active, and it may have been close to a minimum period that happens every 11 yr or so. Solar flares frequently "sweep" the primaries away, but close to the minimum they are rarer, so the number of primaries increases. One of the measurements that shows the increase in the number of events, made using stratospheric balloons in California (USA), is shown in Fig. 12.8.

The other possibility, rather more frightening, is that we are witnessing the inversion of polarity of the Earth's magnetic field, a much rarer process that happens every $(2–3) \times 10^5$ yr. The poles of the Earth exchange their positions and, in the middle of this process, the intensity of the magnetic field goes through a minimum. Thus, the flow of cosmic ray primaries increases, because our "protection" against these particles decreases. It should be pointed out that this has happened countless times in the Earth's geological history and, as we can see, the intensity of the magnetic field has never vanished half way through. But it is worth studying this possibility which, if happens, is irreversible and will mark a new situation for the biosphere we inhabit and for many millennia.

Although we have spoken of the primaries as ordinary particles, there are some very interesting events that suggest that there are some "exotic" ones in their midst that would be worth knowing about. Some of these very different events, which are therefore candidates for having been initiated by exotic primaries, were detected by a Brazil–Japan collaboration in a series of experiments conducted on Mount Chacaltaya, Bolivia, 5000 m above sea level [7]. At these altitudes the rarefaction of

Fig. 12.9 *Left*: The Chacaltaya laboratory in Bolivia. During the winter, the snow made it impossible to collect data in the emulsion chambers. Credit: Francesco Zaratti, Atmospheric Physics Laboratory, La Paz, Bolivia. *Right*: The experiment and a Centauro event, so called because it looks like one thing in the upper chamber and another in the lower. The most remarkable thing is that no photons were detected, so a "normal" electromagnetic cascade had to be discarded. Figure adapted from [7, Fig.38]

the atmosphere—expressed by the low column density, or mass per unit area along the line of sight—allows the study of primaries, mainly hadrons, that interact in the upper atmosphere. The experiment is shown schematically in Fig. 12.9 along with a photo of the setting.

The experiment was composed of two emulsion chambers 158 cm apart. The emulsions were examined with a microscope to determine what kind of particle had passed through them, with the so-called hadronic cascade as the main result. In some events, the expected production seems very different in the upper and lower chambers, so these were called *Centauro events*. In the lower chamber, the primary of the Centauro events did not show the expected range of particle production, and in addition the detected hadrons displayed a huge cross-momentum. The detection threshold of the primaries was of order 1000 TeV. As there is plenty of evidence that an ordinary hadron initiates an electromagnetic cascade (with abundant gamma photons) and a much more "concentrated" hadronic cascade, the consensus is that Centauro primaries are exotic, i.e., objects that do not correspond to any laboratory observations, or rather that are produced at very high energies in certain collisions. It should be stressed that the last possibility has never been confirmed: no Centauro event has ever been observed in any collision experiment carried out in the laboratory. The most real possibility, yet to be confirmed, is that the primary is a small fragment of quarks and gluons in the deconfined state, so there should be active astrophysical sources of these primaries. For the time being, the nature of these events should be considered unknown.

12.1.2   Ultra-High Energy Regime

By general convention, all events above the "ankle" of Fig. 12.2 have been referred as ultra-high energy cosmic rays (UHECR). The question of the highest possible energy for primaries is certainly one of the most important, since it brings with it a number of implications for the acceleration mechanisms and also for the propagation of the particles, as we will see below. Overall, particles that arrive with energies above 10^{18} eV are a problem of the greatest complexity and the utmost importance.

Primaries at these energies are characterized by a short interaction length in the atmosphere, that is, their interaction with atmospheric nuclei takes place right away in the upper atmosphere, giving rise to two different components. The first is an electromagnetic cascade (also called an *air shower*), where the primaries produce a series of secondary particles, which in turn produce a third generation, and so on, until the energy in each particle is no longer sufficient to continue this process. The other component is a hadronic cascade, in a similar process but where the "daughter" particles are hadrons produced by strong interactions. The geometric and energetic development of these cascades can be studied to reconstruct the nature of the primary, its direction of arrival, and its energy, and can be understood using a simple model that we will present below.

The model, due to Heitler [1], assumes a simple type of decay in pairs for each level and provides a rather illustrative analytical treatment. The basic diagrams are presented in Fig. 12.11. For the electromagnetic cascade, each time the particles travel a column length of $X_{EM} \approx 37.6\,\mathrm{g\,cm}^{-2}$, a pair of particles is produced by bremsstrahlung or pair production. This bifurcation continues until the inherited particles have a minimum energy of $E_{min} \approx 86$ MeV, after which they only lose energy without producing new pairs. Thus, after $n = X/X_{EM}$ bifurcations, the number of particles in the shower is $N = 2^n$, since it is a geometric series by hypothesis. At the position X_{max}, all particles reach the minimum energy E_{min}, and the energy E_0 that the primary brought is distributed in its descendants, satisfying $N_{max} = E_0/E_{min}$. Thus,

$$X_{max} \approx X_{EM} \frac{\ln(E_0/E_{min})}{\ln 2} . \tag{12.17}$$

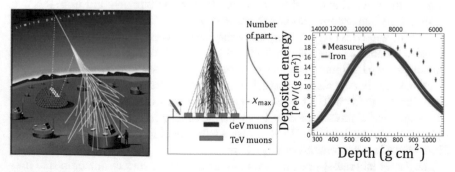

Fig. 12.10 *Left*: An ultra-high energy primary interacts with a nucleus in the upper atmosphere, producing air showers. *Center*: Spatial development of the shower, which reaches maximum particle production for a certain value X_{max}. *Right*: Measurements of events (*points*) and comparison with a set of detailed simulations (much more accurate than the Heitler model, in the *blue band*), where the primary is supposed to be an Fe nucleus, showing the disagreement for this specific case. Figures from [8]

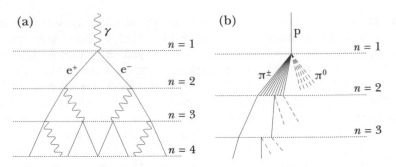

This expression can be used to obtain E_0 by measuring X_{max}. As it is clear that the primary must be some kind of hadron, we see that the electromagnetic shower still dominates the energy balance by far, since the incident hadron produces many π^0 that give rise to gamma photons. Around 90% of the energy goes to the electromagnetic cascade and 10% to the hadronic cascade, where the same reasoning can be applied with the result

$$X_{max} \approx X_0 + X_{EM} \ln \frac{E_0}{\langle n(E) \rangle} \,. \tag{12.18}$$

Here, $\langle n(E) \rangle$ is the average number of secondary particles and each of them carries a fraction $E/\langle n(E) \rangle$ of the energy. An example of the comparison between (12.18) and an actual event has already been shown in Fig. 12.10 (right panel). Of course, the simulations actually used are much more complex and include many important effects and corrections, but the Heitler model serves to show the essence of this procedure.

Having a better appreciation of the showers produced by the ultra-high energy primaries, we can move on to the question of their actual detection. We have already seen that their flux is extremely low—around 1 particle/km² yr—and from this data we know only that experiments must have an enormous effective area for detection, of the order of hundreds of km². To measure the electromagnetic and hadronic showers, we must be able to determine trajectories of secondary particles accurately. The contemporary paradigm of this type of experiments is the Pierre Auger Observatory near the Andes in Argentina (Fig. 12.12). Another important facility is the Telescope Array Project in Utah (USA), an observatory that uses over 500 distributed plastic scintillators and fluorescence detectors with the same purpose of investigating ultra-high energy primaries [10]. We will discuss the specific case of the Pierre Auger Observatory in what remains of this Chapter.

The Auger laboratory operates a very large number of surface detectors and fluorescence detectors spread over an area of about 50×50 km². The surface detectors are plastic tanks, each containing around 12 ton of pure water, together with the relevant electronics and communications antenna to time the events. The tanks are used for detection of Čerenkov radiation (Chap. 2) from the passage of muons in the electromagnetic cascade. These are called the "penetrating component" because

Fig. 12.12 *Left*: Area covered by the Pierre Auger Observatory, near the city of Malargüe in Argentina. *Right*: One of the plastic water tanks that constitutes the arrangement of surface detectors. Some guests in the image will later contribute to the fame of the Argentine barbecue [8]. Credit: Pierre Auger Observatory

Fig. 12.13 Fluorescence detectors (top view). After being focused by a curved mirror, the 6 optical modules (cameras) capture the radiation emitted by de-excitation of the excited levels in the passage of the primary through the upper atmosphere [8]. Credit: Pierre Auger Observatory

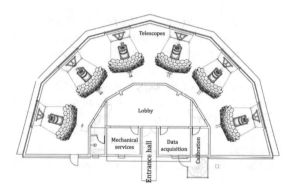

of the low cross-section of the muons, which allows them to reach ground level. Detection operates all the time, with a 100% duty cycle. Fluorescence detectors, on the other hand, observe the tenuous radiation produced by the deposition of energy by showers in the air, with the subsequent de-excitation and production of photons. The device is shown schematically in the form of an optical collector in Fig. 12.13. As it needs good conditions to operate, its duty cycle is just 13% of the total time, corresponding to moonless nights.

These two types of detectors are complementary and their joint measurements provide precise information about events. While surface detectors measure lateral development—i.e., the data set of an event in several detectors allows us to see the cascade developing—relative arrivals in each tank are used to reconstruct the direction of arrival, and finally the energy is determined using a calibrated relationship between X_{max} and the signal at around 1000 m from the extrapolated center of arrival of the primary (there is no substantial error in disregarding greater distances because the energy carried there is too small). However, fluorescence detectors active only

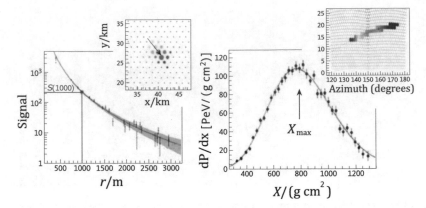

Fig. 12.14 Signal detected by the surface detectors (*left*), which register the shower extension (*upper right*) and the fluorescence detectors (*right*), showing the development that reaches the maximum multiplicity X_{max}. The actual signal is shown in the *upper right corner* [8]. Credit: Pierre Auger Observatory

when there is no Moon and the sky is too dark to see the radiation produced far from them, record the sequential evolution and allow an alternative energy measurement by integrating over the shower profile, i.e., a calorimetric measurement. The recorded signal is shown in Fig. 12.14 for both cases.

Construction of the Pierre Auger Observatory was completed in 2008 and has been accumulating events with energies $E > 1$ EeV $= 10^{18}$ eV, up to the highest energies of around 75 EeV or more, the maximum observed values. In fact one of the biggest enigmas in this area was precisely the existence of this maximum energy. Empirically, the data exhibit a sharp drop in the number of events for $E > 60$ EeV, visible in Fig. 12.15 (see, for example, [11]). This observation is in line with the

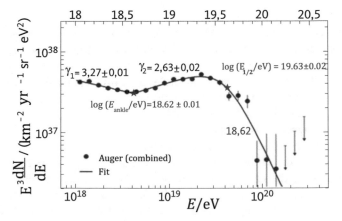

Fig. 12.15 Spectrum of primaries observed at the highest energies. The sudden drop in the number of events above around 6×10^{19} eV, interpreted as the GZK cutoff, is clearly visible (the edge is marked with a *blue star*) [11]. Credit: K.-H. Kampert

theoretical predictions that anticipated a cutoff in the extreme energies, known as the *Greisen–Zatsepin–Kuzmin (GZK) limit*, and which results from the following consideration.

When the primaries of higher energies, supposedly protons, propagate from their distant source, they see the 3 K cosmic microwave background (CMB) radiation as *gamma photons* in their own reference system. The photo-production of pions thus drains their energy quickly and they cannot come from very far away. If the primaries are heavy ions such as Fe rather than protons, things are even worse: the CMB photons would disintegrate them and they would never reach the Earth if injected at the same distance. The relevant reactions in the reference system of the primary proton are

$$p + \gamma_{CMB} \rightarrow \Delta^+ \rightarrow p + \pi^0 \,, \tag{12.19}$$

$$p + \gamma_{CMB} \rightarrow \Delta^+ \rightarrow n + \pi^+ \,. \tag{12.20}$$

Thus, the maximum distance from which a proton can arrive with 60 EeV energy is less than 50 Mpc, regardless of injection energy. This calculation still underestimates the distance because it does not take into account the *multiple* production of pions. The radius of the sphere $R_{GZK} = 50$ Mpc—known as the *GZK sphere*—signals the maximum distance at which the primaries can be injected. Suppression with 20σ statistical significance is assigned to the detected GZK cutoff (Fig. 12.15). However, an alternative explanation is also possible: that the cutoff is unrelated to the GZK sphere and is due to the acceleration mechanism reaching a maximum. There is as yet no definitive exploration of this last hypothesis (see [12] and references therein).

Therefore, the existence of a small number of events with higher energy than the GZK cutoff is also robust. Thus, these events should (a) sometimes originate at smaller distances or (b) be produced by primaries that are *not* hadrons, in order to originate from beyond the GZK sphere and avoid the photo-production of pions. And under any hypothesis, we still have the problem of identifying the sources of these events beyond and below the observed cutoff.

The source of acceleration for these extreme energies is not at all obvious. We have already seen that a sufficient electrical field would be difficult to obtain, due to the tendency of the charges to neutralize potential differences—called *unipolar induction* in the jargon. But the alternative, a first-order Fermi mechanism, also needs an important condition to be satisfied in order to work in this regime: the source must have a size L larger than the Larmor radius r_L, in such a way that the primaries do not escape easily and can reach energies of order 10^{20} eV. As the particles are ultra-relativistic, $E_{max} = p_{max}c$, and this condition $L \geq r_L = p_{max}/Ze\beta$ can be written as

$$E_{max} \approx \beta Z e c B L \,, \tag{12.21}$$

where the product ec is equal to unity in natural units and $\beta < 1$ is the acceleration efficiency when taking into account losses through synchrotron radiation, and so on. Using this expression, one can ask which real systems can produce protons or

Fig. 12.16 Locus of sources that can accelerate primaries up to very high energies as a function of their magnetic field B and size L. Acceleration to the highest energies can happen if the source lies above the diagonal line for a fixed value of β. If the latter is too low, no known source can be a candidate, and even for $\beta \sim 1$ (*dotted line*), there are not many available options. Adapted from [13]

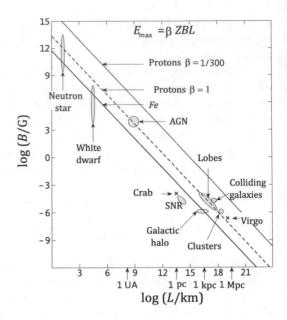

nuclei of a given energy. The result is known as the *Hillas diagram* and is shown in Fig. 12.16.

From the Hillas diagram, we see that the only viable sources in our galaxy, and yet still subject to the efficiency problem, are neutron stars possessing extreme magnetic fields (the magnetars of Chap. 6). It is tempting to conclude that the ultra-high energy regime is due to extragalactic primaries, as anticipated and discussed for nearly a century.

Although the strongest candidates for the accelerating sources are the AGNs from Chap. 8, and the fact that there is a very close AGN, namely the radiogalaxy Cen A at a distance of about 4 Mpc, it has not been possible to confirm that the arrival directions of the higher-energy primaries point statistically to the AGNs (the Auger collaboration announced such a correlation, but this result was not subsequently confirmed). The Cen A lobes are about 60 kpc in size (see Fig. 8.3) and it would not be surprising if it were able to contribute a good fraction of the detected primaries (see below). More recently, the Auger collaboration published a paper [14] in which they claim to detect an anisotropy in arrival directions consistent with extragalactic gamma sources known as *starburst galaxies* (many of them in collision, with their position indicated in Fig. 12.16).

Regardless of these considerations, one should question the analysis used to reconstruct events, which includes numerous theoretical ingredients, among them proton–nucleus or nucleus–nucleus interactions for energies much higher than those ever encountered in laboratories. Figure 12.17 shows that, with these interactions, predictions of the number of muons in cascades are greatly underestimated, a result confirmed by the simulation of the more inclined showers, which are totally dominated by muons. The actual number produced is more than 20% higher than that

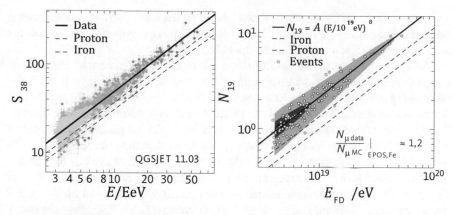

Fig. 12.17 Comparison between simulated proton-initiated (*red dots*) and iron-initiated (*blue dots*) showers for low (*left*) and near-horizontal (*right*) inclinations. The events show an evident lack of agreement. From [16]

indicated by the predictions, and suggests the need for a (difficult) revision of the interactions in the extreme regime.

The Auger data also allow direct investigation of the question of composition through X_{max} and associated quantities. The basic question is: are they protons or are they nuclei? Is it possible to determine their nature? Due to the existence of large fluctuations from event to event, it is most suitable to do statistics with a large set of events for each energy bin, and then compare with the simulations (subject to caveats about the interactions that affect the results, as we just pointed out). Figure 12.18 is the current answer to this question: although close to 1 EeV the composition is compatible with protons, at higher energies the primaries seem to be heavier, viz., α particles and later close to Fe. This is rather surprising and needs to be clarified if we

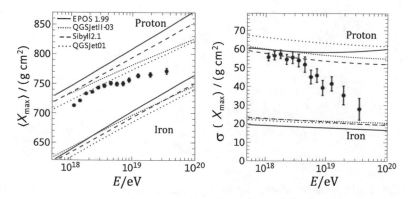

Fig. 12.18 Transition from a "light" composition (protons) to a "heavy" composition (nuclei) at higher energies, consistent with the simulated data in both the X_{max} variable (*left*) and the lateral development (*right*). From [16]

are to progress in this area, since as we have seen from the GZK sphere argument, nuclear primaries can be "brought" from closer distances, even though the sources may be more numerous according to the Hillas diagram.

The last major issue we will address here, which has consequences for the identification of sources, is the question of making "images" using cosmic rays, in the sense of knowing whether the primaries point to the sources that accelerated them. This is what has allowed the development of optical Astronomy, where the neutral photons do not suffer any deviation by the magnetic fields in the intergalactic environment and in the halos of the galaxies. However, we know that the primaries need to be electrically charged (otherwise they could not be accelerated, at least by the mechanisms discussed here). These magnetic fields are uncertain, but polarization data of extragalactic objects show their existence. The intensity is of the order of 10^{-9} G for the intergalactic medium, and up to 10^{-6} G for galactic halos (including our own). The geometry, however, is much more difficult to evaluate. It is precisely for this reason that the arrival of ultra-high energy primaries is an interesting tool, since it has the potential to show deviations in the trajectories that "distort the images" of the sky and sources.

Figure 12.19 shows the paths of the primaries for various energy values, propagating in an intergalactic environment with field intensity $B_{IGM} = 10^{-9}$ G and a correlation length of 1 Mpc. We see that the deviations decrease greatly over the range 10–100 EeV, and the paths there are almost straight. Thus, at the highest energies, the primaries would point to the sources. A previous study has already shown this effect [15], especially if the halo field of our galaxy is ignored. But under any hypothesis, a deviation of 1–2° will remain in the "images" of sources, which are imperfect and in fact slightly distorted.

With the operation of the Auger Observatory and the resulting accumulation of data, it is now a consensus that the observed dipolar anisotropy establishes an

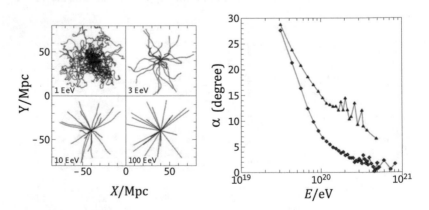

Fig. 12.19 *Left*: Trajectories of ultra-high energy primaries for increasing energies and a fixed intergalactic magnetic field of 10^{-9} G (*left*). *Right*: Angular deviation for the same field, adding 10^{-6} G in the halo (*upper curve*) and a demagnetized halo (*lower curve*), for an extragalactic source at $D = 50$ Mpc [15]

extragalactic origin of particles in the energy range above 8×10^{18} eV [16] since the dipole direction is deviated by 125° from the direction of the galactic center. Thus we have another important piece in the general puzzle.

We have not discussed here other interesting issues, such as the possibility that the primaries are exotic neutral particles of cosmological origin. But it is clear that Nature has produced a "natural laboratory" that no terrestrial acceleration experiment could ever rival, thus providing a unique opportunity to access the world of ultra-high energies. Indeed, it has already given results and promises to be of the utmost importance in helping us to understand the functioning of the Universe.

References

1. T.K. Gaisser, *Cosmic Rays and Particle Physics* (Cambridge University Press, Cambridge, UK, 1991)
2. https://masterclass.icecube.wisc.edu/en/analyses/cosmic-ray-energy-spectrum
3. F. Schröder, *News from Cosmic Ray Air Showers*, ICRC 2019—Cosmic Ray Indirect Rapport. https://arxiv.org/abs/1910.03721
4. M. Bustamante et al., *CERN Latin-American School on High Energy Physics*. http://cern.ch/PhysicSchool/LatAmSchool/2009/Presentations/pDG1.pdf (2009)
5. N. Tsuji et al., Systematic study of acceleration efficiency in young supernova remnants with nonthermal X-ray observations. Astrophys. J. **907**, 117 (2021)
6. M. Longair, *High-Energy Astrophysics* (Cambridge University Press, Cambridge, 2011)
7. C.M. Lattes, Y. Fujimoto, S. Hasegawa, Hadronic interactions of high energy cosmic-ray observed by emulsion chambers. Phys. Repts. **65**, 151 (1980)
8. Home page of the Pierre Auger Observatory. https://www.auger.org/
9. P. Abreu, S. Andringa, F. Diogo, and M.C. Espírito Santo, *Questions and Answers in Extreme Energy Cosmic Rays – a guide to explore the data set of the Pierre Auger Observatory*, Nuclear and Particle Physics Proceedings **273–275**, 1271–1275 (2016)
10. Home page of the Telescope Array collaboration. www.telescopearray.org/
11. K.-H. Kampert, *Proceedings of the 7th International Workshop on Very High Energy Particle Astronomy in 2014 (VHEPA2014)*, JPS Conf. Proc. **15**, 011004 (2017). https://journals.jps.jp/doi/pdf/10.7566/JPSCP.15.011004
12. D. Harari, Ultra-high energy cosmic rays. Phys. Dark Univ. **4**, 23 (2014)
13. P. Bhattacharjee, G. Sigl, Origin and propagation of extremely high energy cosmic rays. Phys. Rept. **327**, 109 (2000)
14. A. Aab et al., (Pierre Auger Collaboration), An indication of anisotropy in arrival directions of ultra-high-energy cosmic rays through comparison to the flux pattern of extragalactic gamma-ray sources. Astrophys. J. Lett. **853**, L29 (2018)
15. G.A. Medina-Tanco, E.M. de Gouveia Dal Pino, and J.E. Horvath, Non-diffusive propagation of ultra high energy cosmic rays. Astropart. Phys. **6**, 337 (1997)
16. D. Gora, The Pierre Auger Observatory: Review of latest results and perspectives. Universe **4**, 128 (2018)

Problems

Selected Problems

The solution of problems is an integral part of the study of any modern science topic. The problems below are a minimal set to encourage understanding and allow the student to gain confidence in the many subjects described in the book. The answers are not given, because slightly different answers will be possible at different levels of approximation and depending on the methodological approach, an intrinsic feature of scientific work that the student should learn to live with. However, in some cases a hint is provided.

1) Taking $k_B = c = \hbar = 1$, convert the following quantities to the natural unit system in powers of MeV:

   a) $T = 8000$ K

   b) $\rho = 2.7 \times 10^{14} \, \text{g cm}^{-3}$

   c) $m = 10$ kg

2) Which conservation law(s) are violated by the reaction $n \rightarrow p + e^-$?

   a) energy

   b) linear momentum

   c) angular momentum

   d) lepton number

   e) electric charge

   f) baryon number

Given that the relevant masses are $m_n = 939.6$ MeV/c^2, $m_p = 938.3$ MeV/c^2, and $m_e = 0.511$ MeV/c^2, write down the correct neutron decay reaction.

3) Explain in some detail the differences between a baryon, a meson, and a lepton. Are they all truly elementary? In what sense?

4) Calculate the Compton wavelength of Schroedinger's cat in units of the Compton wavelength of the electron, assuming its mass is 3 kg.

5) The lifetime of an excited state of a nucleus is of order 10^{-12} s. What is the uncertainty in the energy of a photon from its decay?

© The Editor(s) (if applicable) and The Author(s), under exclusive license to Springer Nature Switzerland AG 2022
J. E. Horvath, *High-Energy Astrophysics*, Undergraduate Lecture Notes in Physics, https://doi.org/10.1007/978-3-030-92159-0

6) Consider a structural model of the electron. Assuming it is composed of particles in a "bag", what should be the energy density of this bag to coincide with the observed size and mass of the electron? What would be the minimum energy of a projectile (external probe) that would be able to "see" this substructure?

7) Show that:

a) A free electron cannot absorb a photon and simultaneously conserve both energy and momentum.

b) A single photon cannot create an electron–positron pair and simultaneously conserve both energy and momentum.

c) These problems do not exist for the Compton effect shown in Fig. 2.2.

8) A pair e^-e^+ is created in such a way that the positron is at rest and the electron has a kinetic energy of 1 MeV, moving in the same direction as the incident photon.

a) Neglecting the energy transferred to the nucleus, calculate the energy of the incident photon.

b) What fraction of the photon momentum is transferred to the nucleus?

9) Describe a setup to check the accuracy of the formula for the Compton effect, in which the visible light from a rubidium laser passes through a plasma at $T = 8000$ K, confined within a spherical region by intense magnetic fields (a Tokamak). Assume that you have a large number of photodiodes and standard laboratory equipment (oscilloscopes, testers, cables, sources, etc.). Draw a diagram to illustrate this setup.

10) Red monochromatic light with wavelength 670 nm produces photoelectrons of a certain metal requiring a stopping potential equivalent to $V_{max}/2$. What is the work function and the threshold wavelength for this metal?

11) How far should we expect X-rays to propagate in the intergalactic medium? Assume the IGM is composed of ionized gas with a mean electron density of $n_e \approx 2 \times 10^{-7}$ cm^{-3}. What is the Thompson scattering mean free path through this plasma? Hint: Use the Thompson cross-section $\sigma_T = 8\pi r_0^2/3$ discussed in Chap. 2. Should we see X-ray sources at distances equal to a significant fraction of the observable universe?

12) An extragalactic source emits jets in which the electrons have an observed Lorentz factor of 10. What instrument would you need to detect photons produced near the source which are assumed to be photons of the 3 K (!) cosmic microwave background upscattered by the inverse Compton effect?

13) The figure below shows a line obtained from an extragalactic source, processed with the appropriate software. The width obtained is $\Delta E = 50$ keV. The astronomer suspects that it comes from the annihilation $e^+e^- \to \gamma\gamma$. What is the temperature of the plasma that produces it if the observed ΔE is attributed to thermal Doppler broadening? What is the likelihood that the emission region is very close to the surface of a neutron star of mass $M = 2M_\odot$ and radius $R = 10$ km?

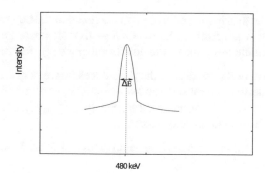

480 keV

14) Search the literature for the most intense synchrotron source in the sky today. How is this emission supposed to arise in this specific case?

15) If we want an angular resolution of $1/10$ of the object size, estimate in how many years we will be able to study the remaining SN1987A in the LMC with optical telescopes that reach $0.1''$

16) Assuming Sedov's expansion precedes the expansion in the *snowplow phase*, find the radius R where this regime changes for a supernova remnant. What is the speed of the ejected material at that radius?

17) Given a power-law distribution of energies of injected particles with index p between E_{min} and E_{max}, in a region with a field intensity B, calculate the synchrotron emissivity and decay time for each energy.

18) A hydrogen plasma from a companion star accretes onto a white dwarf with radius 8000 km and mass $0.5 M_\odot$. The accretion rate is $10^{-9} M_\odot$/yr. The plasma is guided by the magnetic poles of the white dwarf falling onto 1% of the surface. The energy of the material that fell from infinity goes to zero as the material is stopped by a shock just above the surface, so that the material accommodates itself on the surface just below. The region at 1 m depth, just below the shock, effectively absorbs the energy from the material. This region therefore contains a hot plasma, is optically thin, and has a density of 10^5 gcm^{-3}.

   a) What is the number density of ions n_i in the shock region?

   b) Calculate the potential energy lost per second (in erg/s) by the material coming from "infinity".

   c) This energy is converted into thermal energy in the post-shock region. What is the power deposited per cm^3 of this region?

   d) The power radiated by this region is equal to the accreted power in the stationary state. Assuming that all the emission is thermal, calculate the temperature and the emission peak.

   e) Assume now that all the radiated power is bremsstrahlung with Gaunt factor $g_{eff} = 1$. Calculate the equilibrium temperature T of the plasma and the highest emitting band.

19) Suppose that the distribution of protons in the center of the Sun suddenly becomes monoenergetic, with the form $f(E) = A\delta(E - E_*)$. Evaluate the rate of nuclear reactions and calculate the value of the effective exponent β for this new situation.

20) Build a qualitative description of the differences between gravitational collapse supernovas and thermonuclear supernovas. Include in your description comments on the origin of each of these phenomena, the spectral differences, the chemical products that result from them, and so on.

21) Assume that the Sun has a surface temperature of 5700 K, a radius of 1.4×10^{11} cm, and a mass of 2×10^{33} g.

a) Using the Stefan–Boltzmann law, find the rest mass lost per second in the form of radiation.

b) What mass fraction is emitted as radiation every year? How many years can the Sun survive such a loss?

22) An X-ray source moves in the sky with very high, but still undetermined proper motion. The Monomono Satellite observes a line at 3.8 keV that could be assigned to Fe IV (4.1 keV). Discuss what kind of observation you would need to make to determine whether the line is affected by:

a) Doppler effect,
b) gravitational redshift.

23) Find the velocity of a massive object that approaches along the line of sight in such a way that the Doppler effect compensates for the gravitational redshift due to the ratio M/R.

24) The equation of state for a semi-degenerate gas can be written in the form

$$P = K\rho^{5/3}\left[1 + \eta + \eta^2/(1 + \eta)\right] \text{ dyn/cm}^2 ,$$

with $\eta = T/T_\mathrm{F} \approx 3 \times 10^6 T/\rho^{2/3}$.

a) Use the results for polytropes in Chap. 6 to show that the radius of a semi-degenerate object on its way to becoming a white dwarf (after cooling) is $R = R_0\left[1 + \eta + \eta^2/(1 + \eta)\right]$, where $R_0 = 2.8 \times 10^9 (M/M_\odot)^{1/3}$ cm.

b) Invert to express η in terms of the radius, mass, and central temperature of this proto-white dwarf.

25) Consider a spherical shell of finite thickness Δr in a stellar envelope. Assuming that the density is constant inside it and that the value of $\Delta T/T$ across it is large, show that the luminosity of the shell is $L \propto T^4/\bar{\kappa}\rho^{4/3}$.

26) A stellar layer of constant density has a luminosity gradient that grows as the square of the distance Ar^2 from its base to the top. Assume that the opacity $\bar{\kappa}$ is of the Kramers form with coefficient κ_0.

a) What is the dependence of the temperature on r? Assume that the temperature is $T = T_\mathrm{b}$ at the base $r = r_\mathrm{b}$.

b) What is the dependence of the pressure on r?

c) What is the dependence on the rate of nuclear reactions on r, assuming that the layer does not expand or contract?

d) What happens to the temperature variation if the layer (suddenly) becomes convective? Justify, using the appropriate equations for the stellar structure.

27) The Hayashi track of a proto-star on its way to the main sequence is an almost vertical line towards the lower MS, and can be written empirically as

$$\log(L/L_\odot) = 15 \log(T/T_\odot) + 0.2 \log(M/M_\odot) + 8.6 , \qquad \text{(P.1)}$$

while the Henyey track, the final stage before the MS, turning left once the proto-star is radiative and in quasi-equilibrium, takes the form

$$\log(L/L_\odot) = -0.5 \log(R/R_\odot) + 5.5 \log(M/M_\odot) + \text{constant} . \qquad \text{(P.2)}$$

Using the fact that all stars are "black bodies" to a good approximation, obtain the Hayashi track as a function of M and R. Using the fact that, in the lower MS, $R \propto M^{3/7}$, and normalizing with respect to the Sun, calculate the constant in the second equation, imposing the condition that the Henyey track ends (obviously) in ZAMS. Plot (approximately) the two paths in the HR L vs. T diagram.

28) The figure below shows the HR diagram of a $1 M_\odot$ star at the ZAMS. Indicate the following stages on the diagram and complete the statements where necessary:

a) The moment when the core reaches the Schoenberg–Chandrasekhar limit.

b) The Hertzsprung gap region.

c) The path on the giant branch, physically due to ….

d) The position of the *helium flash*, if present. This phenomenon consists of …because of the …condition.

e) The region where stationary burning of helium occurs in the stellar core, known in the literature as ….

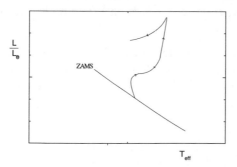

29) To a first approximation, a white dwarf can be treated as a stellar object with constant temperature that doesn't generate luminosity. Integrate the hydrostatic bal-

ance and mass continuity equations for a given density profile $\rho = \rho_0(1 - r/R)$ and construct a family of solutions as a function of the central density ρ_0. Plot the density and determine the radius and the mass of the WD for $\rho_0 = 10^8 \, \mathrm{g \, cm^{-3}}$. Compare with the empirical values published in the literature.

30) The following questions concern the important timescales for the life and death of a star:

- τ_{nuc}

    a) What does τ_{nuc} represent?

    b) Calculate the mass of hydrogen available for fusion throughout the whole life of the Sun. Assuming that 70% of its composition is H, will the whole of that amount be consumed?

    c) What is the physical origin of the Schoenberg–Chandrasekhar limit?

    d) What is the total nuclear energy available, assuming that the mass of the proton is $m_{\mathrm{p}} = 1.67 \times 10^{24}$ g and the mass of the α-particle is $m_\alpha = 6.644 \times 10^{24}$ g?

    e) Using $M_\odot$ and $L_\odot$ to denote the mass and luminosity of the Sun, what is the value of τ_{nuc} for the Sun? Does it coincide with its lifetime on the main sequence? If not, what is the reason for the difference?

- τ_{KH}

    a) What does τ_{KH} represent?

    b) The mathematical expression for this timescale is

$$\tau_{\mathrm{KH}} = \frac{E_{\mathrm{int}}}{L} \approx \frac{|E_{\mathrm{grav}}|}{2L} \approx \frac{GM^2}{rRL} . \tag{P.3}$$

    What approximations have been made to arrive at the last expression?

- τ_{dyn}

    a) What does τ_{dyn} represent?

    b) What events occur in stars on this timescale?

    c) This quantity may be estimated by calculating the time for the stellar envelope to fall onto the center. Carry out such an estimate for the Sun. Hint: recall the Newtonian expression for the gravitational force for the envelope shell.

Comparing the timescales:

    a) What is the hierarchy of these timescales in a stable star?

    b) When the Sun turns into a red giant, its radius will increase to about $200 R_\odot$ and its luminosity to about $3000 L_\odot$. Estimate τ_{KH} and τ_{dyn} for this stage.

c) Indicate the most relevant timescale for each event in the right-hand column of the following table:

Event	Timescale
Contraction pre-MS	
Supernova	
H burning in the core	
He burning in the core	

31) Using the Virial Theorem and the constant density hypothesis, show that the stellar temperature follows approximately the following dependency:

$$T \sim \text{constant} \times M^{2/3}\rho^{1/3} \, . \tag{P.4}$$

If the ignition of any nuclear cycle happens for *fixed* temperature, what can we say about the density at which each cycle for increasing star mass begins?

32) Given a spherical Newtonian star of mass M and radius R, suppose that the density is given by $\rho = \rho_0(1 - r/R)$, that the ideal gas state equation $P = (\gamma - 1)c_V \rho T$ is valid everywhere, and that there is a uniform magnetic field of intensity B throughout the star volume. Calculate the internal energy and show that the virial theorem is satisfied.

33) Determine the equilibrium radius of a white dwarf by minimizing the energy (as shown in Chap. 6) and imposing $E_F \to p_F^2/2m \propto 1/R^2$ for the Fermi energy. Is this radius compatible with observations?

34) Show that the maximum number of electrons that results in a stable structure for a white dwarf (associated with the Chandrasekhar mass) depends only on the fundamental constants G, $\hbar$, and c.

35) Using the approximate numerical value $\xi_1 = 3$ for the zero of the Lane–Emden function, determine the radius of a polytropic model of a white dwarf, given that its index is $n = 1$. Compare with the Earth radius.

36) The density of a toy "neutron star" $\langle\rho_0\rangle = 3M_*/4\pi R_*^3$, with $M_* = 2.8 \times 10^{33}$ g and $R_* = 10^6$ cm, can be treated as independent of the radius to a good approximation. Using this observation as hypothesis, integrate the Newtonian structure equations (hydrostatic balance and mass contained in a sphere of radius r) to obtain the central pressure P_C. Compare the result obtained to the value for more sophisticated stellar models (realistic state equations, general relativity, etc.), which give on average $P_C = 6 \times 10^{34}$ dyn/cm^2. How do you view the result?

37) Use the "cold" Newtonian structure equations to obtain a differential equation for the density as a function of radius for a linear state equation $P = A\rho + B$. Solve it, plot the solutions, and compare with the polytropic solutions already discussed for $n = 5/3$.

38) Briefly explain the concept of the Tolman–Oppenheimer–Volkoff mass. What is its astrophysical application? Hint: Think of the observed masses in Fig. 6.18.

39) The equation of motion of a pulsar that loses energy solely by electromagnetic dipole emission is

$$\frac{d(I\Omega)}{dt} = -KB^2\Omega^3 \, , \qquad (P.5)$$

where I is the moment of inertia, B the magnetic field, and K a constant. Use this equation to estimate the magnetic field of the pulsar with the red circle on the figure below (adapted from [1, Fig. 2]). Hint: If the period P is expressed in seconds and its derivative $\dot{P}$ in $\mathrm{s\,s^{-1}}$, and remembering that $P = 2\pi/\Omega$, we have $\sqrt{I/K} = 10^{15}$ G.

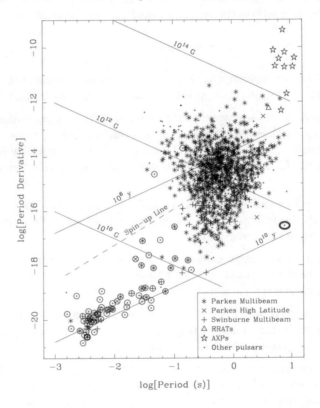

40) Verify that the event horizon area of a black hole is $4\pi R_S^2$. Hint: Remember that the radial coordinate r is *not* the distance to the center. Use the Schwarzschild metric as starting point.

41) The equation

$$v = r\omega = \sqrt{\frac{GM}{R}} \qquad (P.6)$$

describes the speed of a massive particle orbiting a black hole without rotation. However, it can be shown that the orbit is not stable unless $r \geq 3R_S$. Any disturbance will lead the particle in a smaller orbit to spiral down to the event horizon.

a) Find the speed of a particle in the smallest stable orbit around a $10M_\odot$ black hole.

b) Find the orbital period for the same orbit. Compare with the "year" of the planet Mercury, which lasts 88 days.

42) A neutron star with period $P = 1$ s has mass $M = 1.4M_\odot$, constant density, and radius $R = 10$ km. The neutron star is accreting mass from a binary companion through an accretion disk at a rate $\dot{M} = 10^{-9} M_\odot$ yr^{-1}. Suppose the material is in a circular Keplerian orbit around the neutron star until the moment it reaches the surface, and that at this moment all the angular momentum of the material is transferred to the neutron star.

a) Write down a differential equation for $\dot{P}$, the rate at which the period of the neutron star decreases.

b) Solve the equation to find how long it takes to reach $P = 1$ ms, which is about the maximum rotation rate of a neutron star.

43) An accreting compact object of mass M is radiating at the Eddington luminosity corresponding to that mass. An astronaut wearing a white space suit is at rest at an arbitrary distance from the compact object. Assuming that the projected area of the astronaut's body is $A = 1.5$ m^2, find the maximum mass of the astronaut such that the radiation pressure prevents his/her fall onto the compact object.

44) Suppose the Sun collapsed to the size of a neutron star ($R = 10$ km).

a) Assuming that no mass is lost in the collapse, find the period of rotation of this neutron star.

b) Find the intensity of this neutron star's magnetic field.

Although our Sun will not end its life as a neutron star, this shows that the conservation of angular momentum and magnetic flux can easily produce magnetic fields and pulsar-type rotation speeds, at least in principle.

45) Combining gravitation (G), thermodynamics (k_B), and quantum mechanics ($\hbar$), Stephen Hawking calculated the temperature T_H of a black hole without rotation to be given by

$$k_B T_H = \frac{\hbar c^3}{8\pi G M} = \frac{\hbar c}{4\pi R_S}, \qquad (\text{P.7})$$

where R_S is the Schwarzschild radius.

a) Check that the expression has the correct units.

b) A primordial black hole that formed shortly after the beginning of the universe around 13.7 Gyr ago, with a mass of 1.7×10^{14} g, would now be reaching the end of its life. Calculate the temperature of this PBH.

c) The black body temperature T_H corresponds to approximately what range of the electromagnetic spectrum?

d) What would be the radius of a sphere with the density of water if it had a mass of 1.7×10^{14} g?

e) Calculate the temperature of a black hole with mass $10 M_\odot$.

46) Consider Fig. 7.4 showing a binary system with a black hole as primary and $q \neq 1$. If the semi-axis of the orbit is equal to the distance Sun–Mercury and the Roche lobe is $1/4$ of the latter, calculate the mass of the black hole with the aid of Eggleton's fit, assuming a giant companion with $M = 1 M_\odot$.

47) Calculate the fraction f of the bulge mass M_{bulge} carried by the radiation in an AGN with a measured velocity dispersion of 300 km/s which contains a black hole of $10^8 M_\odot$.

48) Calculate how many tons of a certain material a tank should contain in order to detect the monoenergetic neutrinos of the reaction $p + e + p \rightarrow d + \nu_e$ with $E_\nu = 1.44$ MeV, if the cross-section of the material is $\sigma = 10^{-44} (E_\nu/0.5 \text{ MeV})^2 \text{ cm}^2$, to record at least 1 event per month.

49) A type II supernova explodes in the Virgo cluster. Estimate the luminosity expected from the neutrino emission. With the same detectors active at the time of the SN1987A, how many neutrinos would you expect to detect?

50) Suppose there are two types of GRB source with energies E_1 and E_2. Show that if the sources are homogeneously distributed in the universe with numerical densities n_1 and n_2, respectively, then the total number of bursts observed with fluence S is proportional to $S^{-3/2}$.

51) GRBs are classified as "long" or "short" depending on how long they last. What causes are invoked by astronomers to justify this distinction? Comment on the instruments need to detect them and verify their origins.

52) A viable density profile for a white dwarf is given by the expression

$$\rho(r) = \rho_c \left[1 - \left(\frac{r}{R} \right)^2 \right],$$ (P.8)

where R is the radius.

a) Find the total mass $M(R)$ of the star.
b) Find the $M–R$ relation by varying the central densities ρ_c within reasonable values.
c) Show that the average density inside the star is $\bar{\rho} = 0.4 \rho_c$.

53) The contraction of the Sun towards the MS occurred almost in hydrostatic equilibrium. If its initial central temperature was 30 000 K and the final value 6×10^6 K, find the total energy radiated away using the virial theorem. Assuming that the luminosity remained close to the present value of $L_\odot$ (a rather improbable hypothesis), calculate the time to complete the contraction process and reach the MS.

54) When is electron degeneracy important for MS stars? Is it more important in the upper or lower main sequence? Why do ions never reach the degeneracy condition, even in white dwarf interiors?

55) Imagine a gas composed exclusively of electrons. By what process would this gas radiate? What would happen if an equal number of positrons were present? How would you distinguish the two cases in equilibrium at the same temperature?

56) An ultra-relativistic electron is injected in a region with a uniform magnetic field. How does its energy change with time? What happens when the electron reaches the Newtonian limit?

57) A source shows a spectrum that fits $\nu^{5/2}$ at low frequencies, shifts to ν^2 and reaches a peak, dropping as $\nu^{-0.75}$ after that. Is this spectrum consistent with synchrotron emission by relativistic electrons? How do you interpret each behavior?

58) Consider a black hole accreting mass with a fixed efficiency η and always radiating at the Eddington limit. Show that the growth would be exponential and that the timescale is determined by atomic constants alone. (This monstrous behavior is one of the "fears" mentioned in Chap. 8.)

59) Show that N point masses of mass m in the gravitational field of a massive object of mass M achieve a (local) minimum of the total energy when the masses are all in the same circular orbit. Hint: Think of using Lagrangian multipliers.

60) The massive multiple star η Carina may be radiating at its Eddington limit. Estimate the mass needed to explain the observed luminosity of $5 \times 10^6 L_\odot$.

a) In 1837, a "great eruption" happened and η Carinae reached $m_V \sim 0$. Assuming an interstellar extinction of 1.7 mag, without any bolometric correction, estimate the luminosity during this eruption.

b) Calculate the total energy released in photons to sustain the great eruption for around 20 yr, as observed.

c) Read the appraisal by Hirai et al. arXiv:2011.12434 (2020) for some tentative conclusions regarding the physical origin of this eruption and the Carinae system.

61) The flux of neutrinos from the supernova SN1987A was measured to be about 1.3×10^{10} cm^{-2}.

a) If the average energy per neutrino was 4 MeV, use these numbers to calculate the total energy released in neutrinos in this event.

b) Estimate how many human beings acted as a "detector", capturing 1 neutrino from the SN1987A. Hint: Consider the cross-sections of Chap. 9 and approximate an average person by a mass of pure water.

62) Equate the pressure of an ideal electron gas to the degeneracy pressure to find the conditions for the onset of electron degeneracy. Repeat for an ultra-relativistic gas with $v = c$ and compare.

63) Find the size of the Moon if it suddenly converts into a white dwarf.

64) Determine the shortest period of a pulsar in the Newtonian approximation, keeping the fluid spherical and setting $R = 10$ km (shaky approximations indeed). Compare with the period of the fastest known pulsar PSR J17482446ad, for which $f = 716$ Hz.

65) Describe what would happen to the orbits of the planets if the Sun suddenly collapsed to a black hole.

66) a) Could the Sun actually be a black hole radiating Hawking emission? What would its mass have to be for the Hawking emission peak to coincide with the Sun's emission peak?

b) If this were the case, what would the "gravitational constant" G have to be for the Earth's orbit to remain the same?

c) Estimate your own weight on the Earth's surface with this new value of G.

67) In some binary systems, the conditions are such that the angular momentum can be considered to be conserved. Use this approximation to show that the rate of change of the orbital period is

$$\frac{1}{P}\frac{dP}{dt} = 3\dot{M}_1 \frac{M_1 - M_2}{M_1 M_2}. \tag{P.9}$$

68) Consider a photosphere that is being carried away by a shell travelling with velocity v.

a) Show that the density of the photosphere at a distance r is $\rho = \dot{M}_{ej}/4\pi r^2 v$.

b) Within the approximation of a constant mean opacity $\bar{\kappa}$ inside the expanding shell with outer radius R at $t = 0$, show that if the radius of the photosphere (with optical depth $\tau = 2/3$) was R_0, then

$$\frac{1}{R} = \frac{1}{R_0} - \frac{1}{R_\infty}, \tag{P.10}$$

with $R_\infty = 3\bar{\kappa}\dot{M}_{ej}/8\pi v$.

c) At a later time t, when the outer radius of the shell is $R + vt$ and the radius of the photosphere is $R_{ph}(t)$, show that

$$\frac{1}{R + vt} = \frac{1}{R_{ph}(t)} - \frac{1}{R_\infty}. \tag{P.11}$$

d) Using b) and c), obtain the radius of the photosphere as

$$R_{ph}(t) = R_0 + \frac{vt(1 - R_0/R_\infty)^2}{1 + (vt/R_\infty)(1 - R_0/R_\infty)}. \tag{P.12}$$

e) Argue in favor of discarding the terms containing R_0/R_∞ to obtain

$$R_{ph}(t) \approx \frac{vt}{1 + vt/R_\infty}. \tag{P.13}$$

f) Apply this expression to understand the nova explosion shown in Fig. 7.9.

69) Find the photosphere temperature of a nova explosion assuming the Eddington luminosity limit for the event.

70) The *Alfvén radius* is defined as the location at which the magnetic energy density equals the kinetic energy density of the falling material, stopping the fall of matter onto a magnetized object (Chap. 7). Show that the Alfvén radius can be written as

$$R_{\text{Alf}} = 1800 \left(\frac{R}{10\,\text{km}} \right)^{12/7} \left(\frac{B}{10^{12}\,\text{G}} \right)^{4/7} \left(\frac{M}{1.4 M_\odot} \right)^{-1/7} \left(\frac{\dot{M}}{10^{-7} M_\odot \,\text{yr}^{-1}} \right)^{-2/7} \text{km} , \tag{P.14}$$

and therefore that the accreted matter is funneled by the magnetic field far away from the surface for a moderately magnetized neutron star.

71) The AGN population is believed to have been dim throughout cosmic history. Assuming that the number of quasars has been constant since the observed maximum and that the average luminosity obeys

$$L = L_0 (1 + z)^\alpha , \tag{P.15}$$

with L_0 the average luminosity today, i.e., $L_0 \equiv L(z = 0)$, and assuming also that $\alpha \approx 2$, find how luminous a quasar at $z = 2.2$ was (Fig. 8.6).

72) A blazar is an AGN which is thought to point a jet towards the Earth. If the redshift of a target blazar is z_B and the redshift of the ejecta is z_{ej}, show that the speed of the ejecta relative to the blazar itself is given by

$$\frac{v}{c} = \frac{(1 + z_B)^2 - (1 + z_{ej})^2}{(1 + z_B)^2 + (1 + z_{ej})^2} . \tag{P.16}$$

73) Show that when a "normal" object (star, asteroid, etc.) approaches a black hole, the Roche limit beyond which it will be disrupted by the gravitational field is

$$r_R = 2.4 \left(\frac{\bar{\rho}_{\text{BH}}}{\bar{\rho}_{\text{normal}}} \right)^{1/3} R_S . \tag{P.17}$$

a) Using the "average" density of a supermassive black hole discussed in Chap. 6, find the mass of the black hole that would disrupt the Earth at $r_R = R_S$

b) If the Sun is heading towards a supermassive black hole, what is the maximum mass the latter can have for the Sun to be disrupted before entering the event horizon? How does this mass compare with the actual masses in AGNs?

Reference

1. V.M. Kaspi, Grand unification of neutron stars. PNAS **107**(16), 7147–7152 (2010)

Index

A

Absolute magnitude, 60
Accelerating universe, 104
Accretion disk, 165, 167, 169, 171–174
Accretion-Induced Collapse (AIC), 114, 144
Active Galactic Nuclei (AGN), 178, 179,
    181–185
Adiabatic gradient, 70, 71
Afterglow, 227–231
Air shower, 247
Alfvén radius, 169, 170
α-disks, 167
Ankle, 238, 239, 245, 246
Anomalous X-ray Pulsars (AXP), 142, 143
Astrophysical factor, 65
Asymptotic freedom, 10, 12
Asymptotic Giant Branch (AGB), 81
Atomic theory, 2

B

Band parametrization, 224—226
Baryon, 8, 10
BATSE catalog, 225
Baym–Bethe–Pethick equation of state, 132,
    135
Beaming, 228, 231
Beam splitter, 213
BeppoSAX mission, 225
β decay, 7
Bethe–Johnson equation of state, 132
Binding energy, 63
Black body radiation, 19, 29
Black hole merger, 217
Black widows, 171–173
Blanketing, 124
Blazars (BL Lac), 181, 182

Bondi–Hoyle accretion, 165
BOREXINO collaboration, 192
Bound–free process, 23, 69
Bragg condition, 51
Brazil-Japan collaboration, 245
Bremsstrahlung radiation, 30, 33
Broad lines, 177, 181
Bulges, 184, 185

C

Carter diagram, 146
Cataclysmic variables, 170
Centauro events, 246
Central potential, 63, 64
Čerenkov cone, 39
Čerenkov radiation, 38–40
Chandrasekhar limit, 119, 120, 126
Chandrasekhar mass, 92, 97
Chapman–Jouguet adiabat, 99
Characteristic age, 140, 142, 144
Charge-Coupled Device (CCD), 48–50
Chirp mass, 216, 218
Chlorine, 188
Circularization radius, 166
Classical radius of the electron, 24
CNO cycle, 67, 68, 74, 75, 83
Coded mask, 53
Coherent emission, 29
Collecting area, 47, 53, 55
Common envelope, 164
Compactness, 227
Complete Fermi integrals, 199
Compton effect, 24, 27
COMPTON Observatory, 223
Compton wavelength, 6
Continuous sources, 207

Convection, 70, 71, 75
Cooling time, 33
Corotation radius, 169
Cosmic Microwave Background (CMB), 251
Cosmic rays, 237–240, 245, 254
Coupling constant, 6, 7
Crab nebula, 37
Crab pulsar, 97
Cross-section, 22, 24–27
Crystallization, 126, 129
Cumulative gains, 241
Curvature radiation, 38
Cyclotron frequency, 35
Cyg X-1, 156, 157

D
Dark matter, 33, 41, 42
Dark stars, 145
Davies experiment Homestake, 191
De Broglie wavelength, 65
Deep Underground Neutrino Experiment (DUNE), 196, 201
Deflagrations, 100, 101
Deflagration-to-Detonation Transition (DDT), 101
Degenerate regime, 79
Detached systems, 164
Detonations, 99–101
Diffusion, 69, 70, 73
Diffusion time, 240
Dimensionless amplitude, 207, 208
Direct acceleration, 241
Dispersion measure, 233
Dispersion relations, 39
Doppler effect, 41
Double-degenerate, 97, 98, 102, 105
Dynamical timescale, 73
Dynamic range, 49

E
Eddington luminosity, 168
Effective potential, 162, 163
Effective temperature, 59, 72, 77
Electrodes, 48
Electromagnetic cascade, 246–248
Electromagnetic interactions, 6, 7, 13, 16
Electron capture, 92, 105, 106
Elementary particle, 1, 4, 8, 14
Emission cone, 228
Equation of state, 71
Equivalent isotropic energy, 228

Event horizon, 145, 146, 150, 151, 154
Event Horizon Telescope (EHT), 151, 152
Evolved Laser Interferometer Space Antenna (eLISA), 214

F
Fabry–Perot, 213
Fanarhoff–Riley galaxies, 186
Fast Radio Bursts (FRBs), 232–234
Fermi level, 198
FERMI satellite, 219
Fermi theory, 3, 14–16
Feynman diagrams, 15
Fireball model, 228–230
First order Fermi mechanism, 243, 244
Fluorescence, 22
Flux, 19, 29, 30
Focus, 50, 51, 53
Formation of the heavy elements, 219
Free expansion phase, 108
Free–free process, 69

G
Galaxy clusters, 33
GALLEX collaboration, 191
Gamma-Ray Bursts (GRBs), 223, 224, 226–234
Gamow peak, 65, 66
Gauge theories, 15, 16
Gaunt factor, 33
Giant branch, 78–81, 85
Giants, 61, 78
Globular cluster, 123, 144
Gravitation, 7, 8, 13, 14
Gravitational pressure, 134
Gravitational redshift, 41, 42
Gravitational waves, 204, 206, 208, 210–213, 215–217, 219
Graviton, 8, 13, 14
Grazing incidence, 50
Greisen–Zatsepin–Kuzmin (GZK) cutoff, 250, 251
GW150914, 215–218
GW170817, 217, 219

H
Hamuy–Phillips calibration, 103
Hawking radiation, 147, 148
Heavy water, 192, 193
Heitler cascade model, 248
Helicity, 17

Helium flash, 79, 85
High-Mass X-Ray Binaries (HMXB), 171–173
Hillas diagram, 252, 254
Historical supernovae, 89, 91
Homologous collapse, 93
Horizontal Branch (HB), 80, 81
HR diagram, 60, 61, 69, 75, 77, 78, 83, 85
Hybrid neutron stars, 137
Hydrostatic equilibrium, 61, 73, 76, 83, 85, 86
Hypernova, 230

I
IceCube Neutrino Observatory, 201
Infrared slavery, 10
Initial mass function, 113, 114
Injected spectrum, 240
Inner crust, 127
Innermost Stable Circular Orbit (ISCO), 180
INTEGRAL satellite, 218
Interaction range, 6, 16
Interferometers, 210–216, 218
Inverse beta decay, 187
Inverse Compton effect, 24, 25
Iron peak, 83
Isotropic distribution, 224, 225

J
Jetted AGNs, 183

K
K_α iron line, 156
KamLAND, 195
Kelvin–Helmholtz, 74, 77
Kilonova, 219
Kinetic theory, 189
Klein–Nishina, 24, 25
Knee, 238, 239, 244, 245
Kramers form, 69

L
Lagrangian points, 163
Landau–Darrieus instability, 101
Lane–Emden equation, 117, 118, 120
Lanthanides, 219
Larmor formula, 30, 35
Larmor radius, 239, 244, 251
Lepton, 8, 9
Light curve, 45, 46

Light curve SNIa, 100, 102, 103
LIGO observatory, 212
Linear attenuation coefficient, 27
Lorentz force, 34
Lower main sequence, 74
Low-Mass X-Ray Binaries (LMXB), 171–173
Luminosity, 30
Lyman-α forest, 231

M
Magnetar, 105, 107, 110
Magnetic reconnection, 241
Magnetosphere, 140, 142
Mass function, 132, 138
Mass loss, 126
Mean free path, 69
Meson, 8–10, 13
Mestel cooling, 128
Microquasar, 157
Minkowski tensor, 205
MIT bag model, 10
Mixing-length theory, 71
Multimessenger, 55

N
Narrow lines, 181, 182
Negative specific heat, 73
Neutrino, 7–9, 15–17
Neutrino oscillations, 195
Neutrino revival, 95
Neutrinosphere, 95, 199, 200
Neutrino trapping density, 93
Neutron drip, 131
Neutronization, 131, 134
Neutron star merger, 215, 219
Non-jetted AGNs, 183
Non-relativistic Fermi gas, 120
Novas, 170
Nuclear statistical equilibrium, 100
Nuclear timescale, 74

O
Onion structure, 83, 84
Opacity, 69, 70, 77
Optical depth, 228
Outer crust, 137

P
Pair instability, 105—107

Pair production, 25–27, 40
Particle flow (wind), 141
Pauli's exclusion principle, 78
Perturbative vacuum, 10
Photodisintegration, 83, 92
Photoelectric effect, 20–22, 27
Photomultiplier, 213
Photon, 19–24, 26–30, 39, 40, 42
Pierre Auger Observatory, 248–250
Pion bump, 244
Pixel, 48, 49
Planck mass, 14
Polarization modes, 205
Polytropic index, 117, 118
Population III, 105, 231
Positrons, 27, 40–42
Potential well, 48, 49
p-p cycle, 67, 68, 80
Propeller, 169
PSR 1913+16, 208, 209, 211
Pulsar, 114, 130, 138–144

Q
QCD phase diagram, 12
Quadrupole moment, 206, 207
Quantum Chromodynamics (QCD), 9, 11, 12, 15
Quantum mechanics, 4, 5, 14
Quark, 3, 7–13
Quasars (QSO), 177–179, 181, 183–185
Quasi-periodic oscillations, 171

R
Rayleigh scattering, 27
Rayleigh–Taylor instability, 101
Redbacks, 172
Red clump, 80, 81
Redshift, 178, 183–185
Reduced mass, 63
Reflectance, 50, 51
Refractive index, 38, 39
Relativistic instability, 122, 131
Relic neutrinos, 189
Resonant masses, 210–212
Reverse shock, 109, 110
Reynolds number, 167
Rhoades–Ruffini limit, 135, 152, 153
Ringdown, 216
Roche lobe, 163, 164, 166
Roche's problem, 161
Rotating Radio Transient Sources (RRATS), 144

Russell–Vogt theorem, 72

S
SAGE collaboration, 191
Saturated image, 48
Scattering, 21, 23, 24
Schoenberg–Chandrasekhar limit, 76, 77, 81, 85
Schwarzschild criterion, 70
Schwarzschild solution, 146
Scintillation, 51, 54
Second order Fermi mechanism, 241
Sedov–Taylor phase, 109, 111
Semi-detached systems, 164, 170
Sensitivity curve, 213–215
Seyfert galaxies, 181
Sgr A*, 147, 151, 155
Shapiro delay, 138
Shock wave, 38
Single-degenerate, 97, 98, 101, 102, 105
Singularity, 146
SN1987A, 91, 96, 199, 200
Snell's reflection law, 50
Snow-plow phase, 110
SNu unit, 90
Soft-Gamma Repeaters (SGR), 142, 143
Solar neutrinos, 190, 191, 193, 195, 200
Solar wind, 239
Sonic point, 93, 94
Spallation, 240
Spatial domain, 45
Spectral break, 228, 229
Spectral class, 60
Spectral domain, 46
Spectral resolution, 47, 49, 50
Standard candles, 104, 231
Standard model, 8, 9
Starburst galaxies, 252
Star colors, 57–59
Starved BH, 183
Stellar black holes, 143
Strong shock conditions, 109
Structure of matter, 3, 8
Sudbury Neutrino Observatory (SNO), 192, 194, 195
Supergiants, 61, 83
Super-Kamiokande, 192–195
Superluminous supernovae, 105
Supermassive black holes, 150, 151, 154, 155
Supernova Unit (SNU), 191
Synchrotron radiation, 34–37

T

Thermal decoupling, 82
Thermal neutrinos, 197
Thermal pulses, 81, 82
Thermal timescale, 73
Thermonuclear supernovae, 97, 100, 104, 106
Thin shell approximation, 109
Thompson limit, 24
Tolman—Oppenheimer–Volkoff equation (TOV), 130
Total emissivity, 31
Transducers, 212
Transparency, 227
Triple-α process, 68, 80, 82
Tunnel effect, 64, 65

U

Ultra-High Energy Cosmic Rays (UHECR), 246
Ultraviolet catastrophe, 19
Uncertainty relations, 4–6, 10, 16
Unified model, 181, 182
Unipolar induction, 251
Upper main sequence, 74

V

Vacuum fluctuations, 5
Variability, 45
VELA satellites, 223
Velocity dispersion, 184, 185
Virgo galaxy cluster, 207, 220
VIRGO observatory, 212
Virial theorem, 72–74, 83
Virtual particles, 5, 6
Viscosity, 166–168

W

Weak interactions, 7, 14–16
Wein's displacement law, 29
White dwarfs, 113–117, 119, 120, 122–131, 134, 136, 138–140, 144
White dwarf seismology, 129
Work function, 21, 22

X

X-ray binaries, 114, 152–154

Y

Yukawa potential, 132

Printed in the United States
by Baker & Taylor Publisher Services